Helicopter Aerodynamics Volume III

Ray Prouty's Vertiflite Magazine* Columns 1990-2013

Eagle Eye Solutions, LLC

Copyright 2016 by Shawn Coyle

ISBN - 978-0-9792638-7-3

Published by:

Eagle Eye Solutions, LLC
P.O. Box 27
Charlotte, VT 05445-0027

Phone   1-661 333 5976
email: shawn.coyle@EagleEyeSolutionsLLC.com
Web   www.EagleEyeSolutionsLLC.com

Prouty, Ray

Helicopter Aerodynamics Volume III

First Edition

Printed in the USA by Lulu.com

Bibliography: p.

1. Books - United States - Aviation   1. Prouty, Ray -2013 -

II Title - Helicopter Aerodynamics Volume III, 2016

III Title - Helicopter Aerodynamics Volume III

FOREWARD

Ray Prouty was one of those rare people who combine technical expertise, the ability to communicate clearly and has had a wide range of unique experiences. His columns in Rotor and Wing starting in 1979 were the first to explain the complexities of helicopters in clear, easy to understand language. These columns were collected into a series of three books in the early 1990s, and they remain treasured volumes by those lucky enough to have them. We reprinted these columns in a single book "Helicopter Aerodynamics Volume I.

Ray stayed busy after 1992, and continues to write for Rotor and Wing and for the American Helicopter Society's magazine "Vertiflite." We've taken those columns and produced this book. It should be noted that the columns were written over a period of several years, but have been arranged here not chronologically, but by subject matter. In some cases, Ray has changed his mind when writing about the same subject and so you might find some small discrepancies between adjoining chapters. Please be understanding.

I knew Ray for quite a few years, and always admired both his extensive (both breadth and depth) technical knowledge and ability to communicate clearly with subtle humor. Ray's technical knowledge was borne of long experience in helicopters, starting as an aeronautical engineering student and culminating in normal "work" as the Chief of Flying Qualities Analysis for Hughes Helicopters / McDonnell Douglas Helicopters.

Lots of other engineers have similar career paths, but few have spread it across as many companies and projects. Even those who come close haven't attempted to write about it. His library was one of the most amazing collections of helicopter books, drawings and notes in private hands. Nearly all the photos in this book came from his collection.

One of the notable things about Ray's experience was that he obviously observed and paid close attention to what was going on around him, and that he kept notes! One of the pilots I spoke to said that Ray was legendary for his filing system - if you wanted to know something about cotter pins, for example, Ray was sure to have several inches of papers filed away under 'Pins, Cotter'. But there's was more to Ray than just filing things away. There was integrating the information. Ray put together information from a variety of sources and made it all understandable. He collects information from technical papers that are full of formula and graphs and translates it into normal English.

There is something else that's difficult to pin down, shining in the background, something elusive. Something rare. Perhaps that "something" is best understood by looking to see if others had done what Ray had done - and the answer to that was that almost no one else had ever put so many of the complicated concepts that make up rotary wing aviation into simple, easy to understand language.

And the helicopter community is much, much the better for his work. It's with great pride Eagle Eye Solutions puts this volume into print. It is with a heavy heart that I have compiled this last book o Ray's columns. Ray passed away after a long and full career in the rotary wing world, filling many roles admirably - engineer, speaker, advocate, teacher, if any of these are separate and easily labeled. Personally, he was always a guiding light on the technicalities of helicopters, and kept me up to speed on these most complex of flying machines. He inspired me to write my books and columns. He will be missed, greatly.

Shawn Coyle

Contents

Contents

Contents

Contents

CHAPTER 1 *Induced Power*

The Fixed Wing People Started it...

The airplane aerodynamicist needs to know how much of his aircraft's drag is due to the rearward tilt of the wing's lift vector. This is the induced drag. The early pioneers in this field used the laws of physics to develop a simple equation which is still used.

One of the features of this equation when it is converted to induced power is that forward speed is in the denominator. This makes the calculated induced power decrease as speed increases. This is because the wing has more air per second to use in developing lift.

The airplane aerodynamicists found that the minimum wing induced drag was achieved when the circulation in the far wake had a semi-elliptical distribution. This could be produced by an elliptically shaped wing. The British Spitfire fighter of World War II could fly a little faster and a little farther because it had an elliptical wing rather than the more conventional shape that most other airplanes had. The other side of the coin is that this wing is harder to design and build than a straight or tapered wing.

To account for the penalty in induced drag for those other airplanes without elliptical wings, the equation for induced drag coefficient is modified by inserting a factor less than unity into the denominator. This is known as the Oswald efficiency factor,

(Bailey Oswald was an aerodynamicist at Douglas Aircraft Company during World War II). The value of e can be estimated by comparing the shape of the wing to that of an equivalent ellipse. Values from 0.7 to 0.9 are common. Note: in some derivations, e is replaced by (1+?).

We Picked It Up

The first rotary-wing aircraft, the autogyro, was just an airplane with a strange wing.

H. Glauert in England, found that he could use modifications of the airplane equations for rotor induced effects. These equations have been passed on to helicopter aerodynamicists.

A Scientific Study

Robert Ormiston in a 2004 AHS Forum paper reports on a computer study of induced power of helicopters using modified equations developed for airplanes. He shows that at tip speed ratios above about 0.3 the induced power stops decreasing and starts increasing and can be several times the hover value at tip speed ratios approaching 1.0! As Ormiston points out, the source of the phenomenon is something airplane people don't have to consider "a reverse flow region producing negative lift. On a helicopter, the size of this region grows as forward speed is increased. It is a function of the tip speed ratio. If this value is 0.5, the region extends half way out on the retreating blade. At a tip speed ratio of 1.0, the entire retreating blade is in it. Since the blades on the retreating side are expected to produce their share of positive lift, it can only be done by those blade elements outside the reverse flow region, in the "normal flow "region, but as speed is increased, that region becomes smaller and smaller until at some forward speed, retreating blade stall is encountered. Even up to this point, the effect is being felt by the rest of the rotor. The decreased ability of the retreating side to produce positive lift means that the lift on the advancing side must be reduced to maintain roll trim. This, in turn, means that to maintain a constant rotor thrust, the lift on the blades over the nose and the tail must be increased. It is their higher induced drags that account for the rise in total induced power.

This does not apply to autogyros which fly with their rotors tilted up rather than down. Their reverse flow regions produce positive lift and therefore their calculated induced powers are similar to airplanes.

Not to Worry

The use of something similar to the Oswald efficiency factor for helicopter induced effects should only apply when calculating rotor power by the momentum method. If a blade-element method is being used, no "fudge factor "is needed. The program will integrate the effects of the tilt of lift on all the blade elements to produce a value of "total induced power "that is accounting for the fact that the rotor is acting both as a wing and as a propeller. The true induced power due only to lift--as on a wing--is found by subtracting the parasite power associated with the drag of the fuselage and other components that account for the need for a propulsive force. (Note: This is not done in all studies of induced power.)

An Example

Figure 1-1 shows the result of calculating induced power for the example helicopter of my textbook. The only change has been to use a more advanced airfoil (the VR-7) than the original NACA 0012 to obtain a higher maximum speed. For bookkeeping purposes, the tail rotor drag and the main rotor H-force due to skin friction have been added to the drag of the airframe. It may be seen that at the maximum speed of 210 knots "a tip speed ratio of 0.54'the calculated induced power is more than three times the value that would be calculated from the simple momentum theory. This agrees with Ormiston's results at this condition.

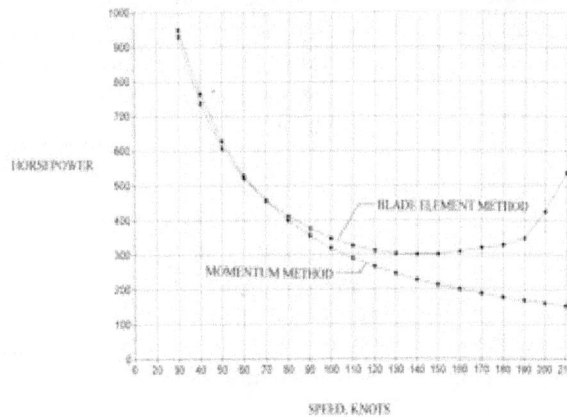

FIGURE 1. Calculated Induced Power for the Example Helicopter

Even the line based on the momentum approach in Figure 1-1 is a little misleading. Its calculation--as in most studies of this type--was based on a rotor thrust being equal to the gross weight of the helicopter. But because Glauert's equation actually has the square of the thrust in the numerator, this should have been accounted for at high speeds where the fuselage drag is high and the fuselage must be tilted down and the rotor thrust must be increased to maintain enough vertical component to balance the gross weight. Using this effect for my helicopter, the increase in thrust would make the calculated momentum induced power begin increasing--instead of decreasing--at about 170 knots.

To the Extreme

Ormiston extended his computer study to tip speed ratios up to 1.35. As they approached 1.0, the computed induced power went divergent as the download on the retreating side overpowered the rest of the rotor and it could not be trimmed.

The Moral is:

Don't rely on only a rotor to hold the helicopter up at very high speeds. Use a wing to unload the rotor and add a propulsive device to make up for the lost ability of the rotor to pull the aircraft through the air. This, of course, results in a compound helicopter in which the rotor carries only a little thrust without running into retreating blade stall.

An Alternative

A possible solution for tip speed ratios above 1.0 is to use the configuration known as the "Reverse Velocity Rotor. "Here, the entire retreating side is in the reverse-flow region and the cyclic pitch is changed twice per revolution instead of only once as on conventional helicopters. With this scheme, the reverse-flow region can produce positive lift, (though with a not-so-efficient-airfoil because it is going backwards through the air) while the advancing side is also producing positive lift. Ormiston shows a reasonable level of induced power for this RVR design, but only beyond a tip speed ratio of about 1.2.

Putting the Computer to Work

I have used my forward flight blade-element program on my example helicopter to investigate the effects of several configuration changes. Starting at 30 knots, the program increases speed in 5 knot increments up to 185 knots. This is at a tip speed ratio just under 0.50 which is generally assumed to be the limit for conventional helicopters. (The Lynx speed record of 216 knots was set at 0.50 with a higher tip speed than my design.)

Case Studies

Changes of various configuration parameters have been investigated. The first was blade twist. The baseline was -10o. The other two were -5o and -15o. For each twist and speed, the ideal induced power, based on the Glauert assumption, was divided by the calculated induced power to get a factor that I have called "e "as an airplane aerodynamicist would. Figure 1-2 shows the result (a high value of e is good, resulting in low induced power.) The plot indicates that low twist is good for reducing induced power in forward flight, but we all know that good hover performance depends upon high twist. This illustrates the dilemma that helicopter designers constantly face: "Whatever helps forward flight, hurts hover and vice-versa "

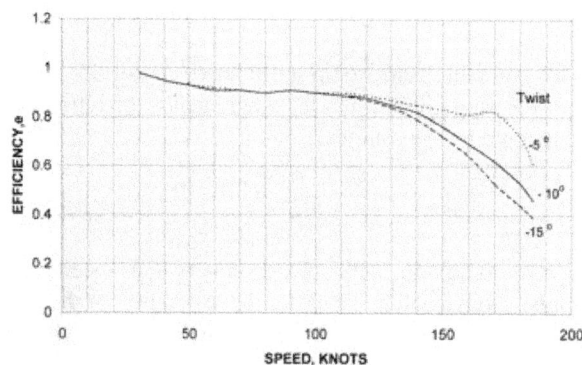

FIGURE 2. Effect of Twist on e

The next study had to do with blade taper. My design has constant-chord blades so I also looked at blades with normal taper with the chord decreasing from root to tip and with reverse taper with the opposite trend. The amount of taper was defined by a number. For normal taper it was 0.5 which means that the tip chord was half the root chord. A value of -0.5 represented the opposite. In each case, the blade chord at the 75% radius station was the same as the untwisted blade: 24 inches. This maintained the same thrust-weighted solidity. Figure 1-3 shows that reverse taper is best. It backs up the argument for the BERP (British Experimental Rotor Program) paddle tip used on the EH-101.

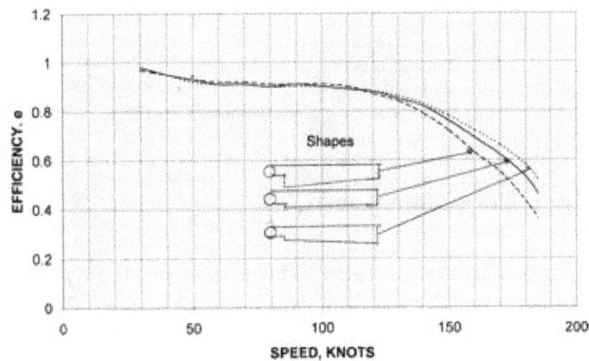

FIGURE 3. **Effect of Taper on e**

Another configuration change that was investigated involved the blade station at which the air-foil actually begins: the "cutout. "The normal value for my helicopter is 0.15 of the radius, so 0.05 and 0.25 were also used. Figure 1-4 shows the results of this investigation. The low value of e for 0.05 cutout indicates that the aerodynamics close to the hub is causing trouble. Starting the airfoil at 0.25 is better. The Sikorsky CH-53E has this much cutout.

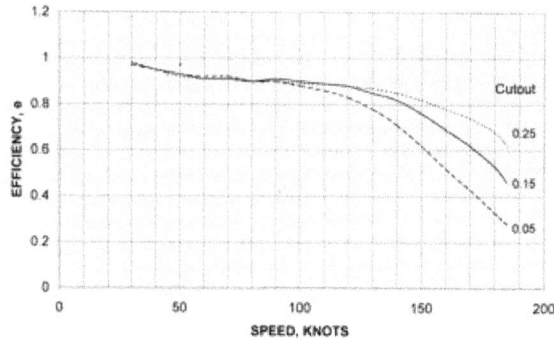

FIGURE 4. Effect of Cutout on e

How about HHC (Higher Harmonic Control) This is done by slightly changing the cyclic pitch twice per-revolution as a modification of the normal once per-revolution change. It can be done with a lateral input, A2, or with a longitudinal input, B2. In the first case, for a positive value of A2, the pitch is reduced by A2 degrees when the blade is over the tail boom and then again when it is over the nose. B2 does the same from side to side. The results are shown in Figure 1-5 for lateral HHC and in Figure 1-6 for longitudinal HHC with each using two degrees in each direction.

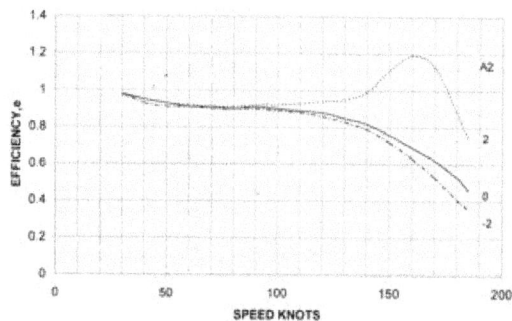

FIGURE 5. Effect of A2 (Lateral Input from HHC) on e

Figure 1-5 shows a surprising effect for the case with positive A2 at high speed. The calculated induced power is less than the ideal induced power! At this point, I can only use the old explanation of computer people, "Well that's what the computer says "

I hope someone will explain this to me someday.

On the other hand, Figure 1-6 indicates that longitudinal cyclic pitch, B2, has little significant effect on induced power.

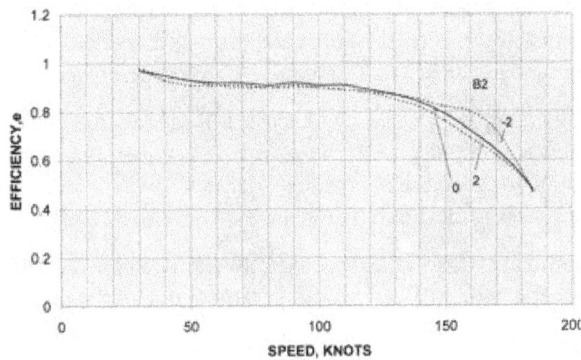

FIGURE 6. Effect of B2 (Longitudinal cyclic input from HHC) on e

The Bottom Line:

I would like to use the information from above to find ways of improving the cruise performance of my helicopter. Since the maximum speed is 185 knots, I will choose a cruise speed of 170. The result is in the table which lists the various configuration parameters and the significant trim results at this speed. Case 1 is the baseline helicopter

Case	Twist	Taper	Cutout	A2	B2	H Force	α Fuse	HPPAR	HPPRO	e	HPIND	HPMR
1	-10.	0.	0.15.	0.	0.	-11.	-9.4.	1419.	686.	0.62.	380.	2485.
2	-5.	0.	0.15.	0.	0.	-281.	-8.6.	1350.	917.	0.82.	281.	2548.
3	-15.	0.	0.15.	0.	0.	233.	-10.	1490.	561.	0.53.	447.	2498.
4	-10.	0.5.	0.15.	0.	0.	51.	-9.6.	1441.	805.	0.55.	427.	2672.
5	-10.	-0.5.	0.15.	0.	0.	-45.	-9.3.	1408.	642.	0.44.	357.	2407.
6	-10.	0.	0.05.	0.	0.	149.	-9.9.	1466.	556.	0.43.	545.	2567.
7	-10.	0.	0.25.	0.	0.	-128.	-9.2.	1396.	1097.	0.74.	315.	2808.
8	-10.	0.	0.15.	2.	0.	-323.	-8.5.	1341.	1058.	1.13.	203.	2602.
9	-10.	0.	0.15.	-2.	0.	147.	-9.9.	1461.	600.	0.53.	450.	2511.
10	-10.	0.	0.15.	0.	2.	-128.	-9.	1387.	689.	0.64.	365.	2441.
11	-10.	0.	0.15.	0.	-2.	-17.	-9.4.	1422.	876.	0.72.	326.	2663.

It may be seen from this table that the induced power plays only a small role in this study. The H-Force is listed because it plays an important role in the comparisons. It is primarily due to the difference in skin friction between the advancing and retreating sides of the rotor. At moderate speeds, the high dynamic pressure on the advancing side is more than on the retreating side and the H-Force will be positive, acting to the rear. For my helicopter, it is 50 pounds at 140 knots. At higher speeds with higher angles of attack required on the retreating side, its higher drag coefficients begin to produce large drag forces that act forward instead of backward. This results in the rear-pointed H-Force being reduced and eventually reversing in sign. For my helicopter, this happens near 170 knots. At 185 knots, much of the retreating side is in stall and the H-Force is -245 pounds. Now the retreating side is acting like an oar in water and is helping to overcome parasite drag. In this situation, the rotor tip path plane need not be tilted as far down as with a positive H-Force and the angle of attack of the fuselage is less negative which converts to lower fuselage drag and less parasite power.

This is illustrated by comparing Case 2 with Case 1. Reducing the twist from -10o to -5o does reduce induced power, and it also reduces parasite power--two good results. But because the rotor is closer to stall, the increase in profile power more than makes up for these reductions and the total rotor power is increased by 63 horsepower. So much for reducing twist!

Another disappointment is Case 7 that predicted low induced power with a 0.25 cutout. This is being sabotaged by the high profile power requirement due to the unstreamlined blade spars.

Similar effects can be deduced from comparing the other cases with the baseline. Based on total rotor power (HPMR), only two changes appear to be worth pursuing: Case 5 with reverse taper, and Case 10 with HHC and positive B2. This latter effect was discovered in flight test of HHC on the Hughes OH-6.

Another Approach

The airplane people, including Max Munk, E. Trefftz, and Ludwig Prandtl developed a sophisticated method for determining the induced power of a wing in forward flight many years ago.

The idea is that the kinetic energy of air motion within the wake left by a wing is related to the induced drag. The kinetic energy, in turn, is related to the circulation within the wake which was produced by changes in lift from one wing element to another as you go from one wing tip to the other. Circulation is a scalar quantity or in other words, it is not a vector. To use a crude analogy, the distribution of circulation in the wake of a wing is similar to the depth of the pile left behind a farmer's manure spreader.

It can be shown that if the distribution of circulation in the wake has the shape of a semi-ellipse, two things will be true: the induced velocity across the wake will be a constant, and the induced drag will be minimum. If the circulation distribution is not elliptical, the extra induced drag increment can be calculated by the mathematical procedure developed by those aerodynamic pioneers.

How about a rotor?

The helicopter rotor also deposits circulation in its wake. The circulation is not only produced by lift changes as a function of the blade-element radius station, but also as the lift changes as a function of blade azimuth. A blade element computer program can be used to calculate these changes and to find the circulation dropped by each blade element which taken together form the wake. So it is just a matter of plotting the spanwise station of the resulting circulation strength in the wake and then comparing this distribution to the corresponding ideal semi-ellipse that would be generated by an ideal rotor lifting the same amount. The semi-ellipse would correspond to the situation usually assumed in the momentum (Glauert) approach for calculating helicopter induced power.

Figure 1-7 shows the distribution of circulation in coefficient form across the wake of my example helicopter at its maximum speed of 210 knots. The distribution is only approximately the shape of the ideal semi-ellipse. Following the method used for wing analysis, the dimension across the wake can be transformed into a axis system where the progression from one side to another is related to half a cycle from 0o to 180o. This would convert a semi-ellipse into half a sine wave.

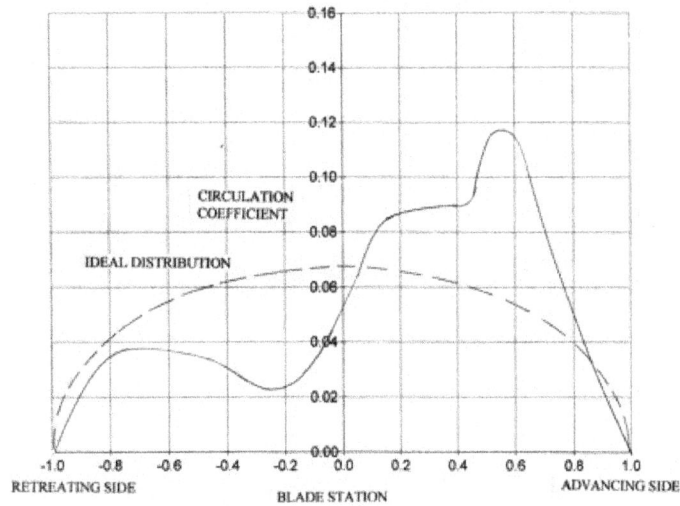

FIGURE 7. Distribution of Circulation on Rotor Wake

This transformation allows the distribution to be expressed as a trigonometric function consisting of the superposition of many frequencies. Each frequency has its own magnitude which is expressed as a Fourier Coefficient. The first Fourier coefficient represents the rotor thrust, and the second represents the rolling moment. All the coefficients are used to find the induced drag.

For this computed distribution, the induced drag for my helicopter at 210 knots is 3.45 times more than what it would be on an ideal rotor with a semi-elliptical distribution. This is essentially the same ratio that can be obtained from the Figure 1-1 at the same speed.

And so What?

This example shows how procedures developed for wing aerodynamics can be applied to rotors, but as stated above, fortunately if we use blade-element methods to predict the performance of a rotor, we do not need to resort to such an involved process.

CHAPTER 2   *The Reverse Flow Region*

Most of the work of the rotor is done in the normal flow region where the air strikes the leading edge of the airfoil. But there is another region which should be considered for its detrimental effects on forward flight performance: the reverse-flow region where the combination of forward speed and rotational speed produces the situation where on part of the disc, the air strikes the trailing edge of the blade.

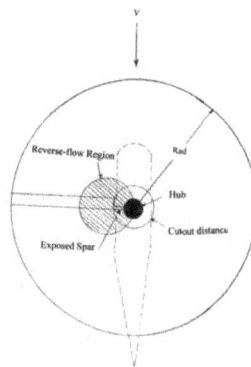

FIGURE 1. Geometry of the Reverse-Flow Region in Forward Flight

The reverse-flow region is always shaped as a circle whose inboard edge is located at the center of rotation and it extends outboard to a blade station on the retreating blade equal to the tip speed ratio. Thus at low forward speeds, it is very small, but at speeds associated with setting helicopter speed records, it may extend half-way out the retreating blade. Figure 2-1 shows the geometry of the rotor of the 20,000 pound helicopter of my textbook at 172 knots. At this speed, the tip speed ratio is 0.45 which means that the reverse-flow region extends 45% out the retreating blade.

The combination of the nose-down tilt of the tip path plane, the collective pitch, the cyclic pitch, and the local downwash produces an angle of attack in this region that is negative when referred to the trailing edge. This produces negative lift and drag vectors that are contributing to the download in this region. Also most of the blade elements here are in a stalled condition since the sharp "leading edge "cannot support good flow conditions as can the normally rounded airfoil nose. This drag in the reverse-flow region requires more engine power as if it were parasite drag.

Figure 2-2 shows the aerodynamic conditions on a blade element near the center of the reverse-flow region.

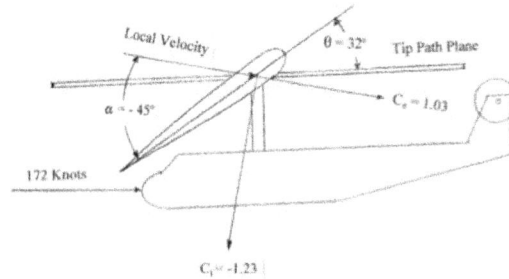

FIGURE 2. Conditions at the Center of the Reverse-Flow Region

At low speeds, the reverse-flow region does not extend much beyond the blade cutout and so little of the rotor disc is affected. Even at higher speeds, the local effects are small near the boundary where the dynamic pressure is zero.

Figure 2-3 shows the calculated distribution of download in terms of pounds per square foot on blade elements at a speed of 172 knots. This download must be compensated for by higher thrust in the normal-flow regions.

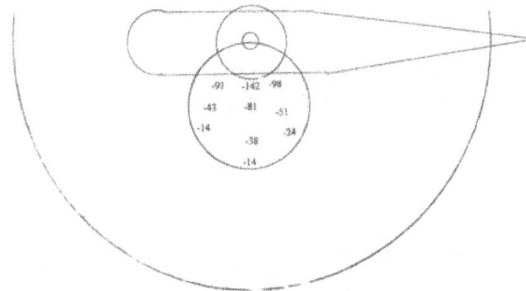

FIGURE 3. Download Distribution in the Reverse-Flow Region, pounds/ sq ft

The integration of the forces on all the elements in this region gives a download of 338 pounds and an H-force of 210 pounds, both of which are hurting performance.

Putting the Computer to Work.

I have investigated these effects on my helicopter by using a blade element performance-and-trim computer program. For a given flight condition of gross weight, speed, and altitude, the program puts the helicopter into trim with respect to the six degrees-of-freedom by solving for the four cockpit control positions, and the pitch and roll attitudes with respect to the horizon.

It then calculates all the conditions of interest such as the various components of required power and H-Force for both the main and tail rotors, and the forces and moments on the rest of the helicopter components. The program increases forward speed in increments until it can no longer find a solution due to retreating blade stall.

For this study, I also had the program keep track separately of conditions within the reverse-flow region to solve for its download, H-Force, and the power associated with induced and profile torque effects.

Each blade is connected to the hub by a cylindrical spar three inches in diameter whose drag is accounted for using a drag coefficient for a cylinder at each increment of span (the program uses 50 blade elements starting at the center of rotation). The spar extends out to where the actual blade starts which is 15% of the 30-foot radius. This value is called the "cutout "and can be changed for trade-off studies.

As a start, I modified the program by eliminating the lift and drag at each blade element in the reverse-flow region. The result can be called "the baseline. "Comparing the computer output of this with the output of the full program gives an indication of how much trouble the reverse-flow region is making - in this case, at 172 knots, the reverse-flow region is accounting for a total power increase of 164 horsepower, or 6% of the total.

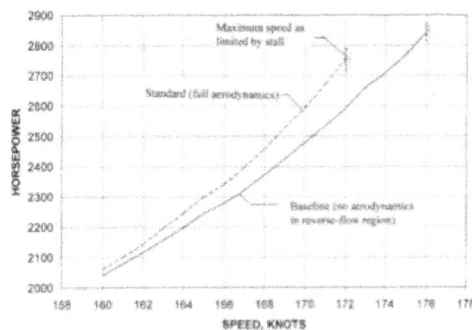

FIGURE 4. Effect of Reverse-Flow Region on Power Required

With the VR-7 airfoil which permits higher speeds than the NACA0012, the reverse-flow losses are even greater as shown in Figure2-5.

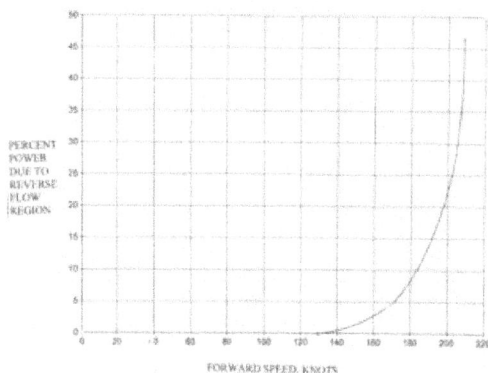

FIGURE 5. **Reverse-Flow Effect at High Speed**

The Cutout Question

Since the forces in the reverse-flow region are due to the aerodynamics of the area of the blade in it, it seems to make sense to increase the cutout to eliminate some of that blade area. This was investigated by running the program with cutout values between the value of 6.7% (where there could be no exposed spar on my helicopter) up to the 25% that you can see on the Sikorsky CH-53E.

The result of this trade-off is shown on Figure 2-6 as the power required at cruise and also the maximum speeds as affected by cutout. At low values there is too much blade area in the

14

reverse flow region resulting in a high power loss, but as the cut-out gets higher, things improve, reaching an optimum (for my design) at about 12%.

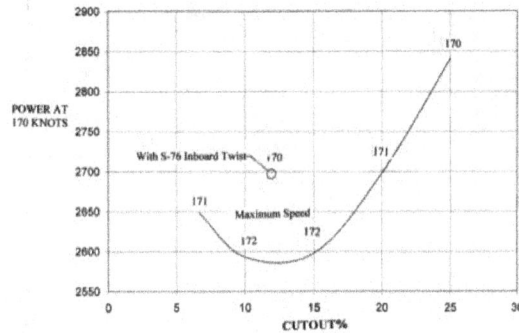

FIGURE 6. **The Effect of Cutout Value**

Increasing cutout higher does reduce the area in the region, but it also increases the exposure of the un-streamlined blade spars at all the other azimuths with a resulting power penalty. Thus the two effects tend to counter each other. At high cutout, we are also losing some blade area that we had before. This effects retreating blade stall and accounts for the reduction of maximum speed.

The large cutout on the Sikorsky CH-53E came about when this helicopter was being developed from earlier versions. As the third engine was installed and the gross weight increased, it was found desirable to increase the rotor diameter from 72 feet to 79 feet (the number of blades was also raised from six to seven). It was also decided to use the same blade design as on the earlier versions. The change was done by installing a 3 ½ foot long spar extension at the root of each blade. My calculations would indicate that Sikorsky is paying a significance performance penalty for this decision.

The Sikorsky S-76 twist

Another possibility is to modify the inboard twist distribution as on the Sikorsky S-76 shown in Figure 2-7. The modification was based on the same reasons given above and was discussed in a 1978 AHS Forum paper by Fradenburgh on the development of the S-76. His discussion includes the statement, "Model tests of the blade root area had previously shown that the drag of this region could be significantly reduced by such a local untwisting of the blade, combined with proper fairing of the blade trailing edge. "Unfortunately, Fradenburgh did not tell us about that "proper fairing "

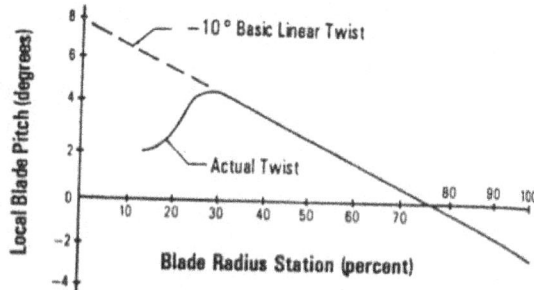

FIGURE 7. Blade twist distribution on Sikorsky S-76

My calculations show that the modification has a slight advantage up to 60 knots, but faster than that, the power required is higher than with the standard linear twist. At high speed, its performance is plotted as a point on Figure 2-6. The detriment in performance is apparently due to changes in trim conditions that increase the angles of attack on the retreating blade. Perhaps the fact that Sikorsky doesn't use this on their other helicopters tells us something.

CHAPTER 3   *A Windfall for Vertical Climb*

To insure that a helicopter can do the maneuvers that might be needed under combat conditions, the US Army required that both the Sikorsky Black Hawk and the Hughes (Boeing) Apache should have the capability of making a vertical climb of 450 feet per minute at an altitude of 4000 feet and a temperature of 95o F while using only 95% of their available power. The question is: How much additional power is needed above that required to hover out of ground effect?

A Simple Calculation?

From the laws of basic physics, this should be an easy calculation. Just find out what the rate of increase in potential energy is by multiplying the gross weight by the climb rate and then dividing the result by 33,000 to get the additional horsepower. Simple but wrong!

This would be right for an elevator, but the helicopter has something else going for it. The induced power is a function of the mass flow of air through the rotor disc. While increasing speed in vertical climb, the amount of air the rotor has to work with increases and the rotor doesn't have to work so hard on it. Thus there is a "windfall."

Using some assumptions and the same momentum methods we use for hover analysis, it is easy to show - at least for a climb rate that is small with respect to the hover induced velocity - that the additional power increase is only half of what would be required to raise a similarly loaded elevator at the same rate.

OH, Those other Effects!

This calculation is usually close enough for a first estimate but, as usual in helicopter work, there are a number of "yes, buts. "These involve three factors:

1. The fuselage download is increased by the increased velocity passing past it.
2. The tail rotor power must be increased to balance the increased main rotor torque.
3. The profile power due to the distribution of angles of attack on the blades will change from the hovering condition.

Correcting the first approximation for these effects as applied to my example helicopter, the increased power to climb at 450 feet per minute is not 50% of what an elevator would need, but is 75%.

More Complication

We're not through, however. The above statements are based on the assumption that the rate-of- climb is small compared to the induced velocity. For low (up to 500 feet per minute) rates, this assumption is "good enough for government work', but if you have a helicopter that climbs like a rocket, you should use a more complicated equation in which the rate-of-climb is combined with the induced velocity to get the mass flow through the rotor. Figure 3-1 shows the difference.

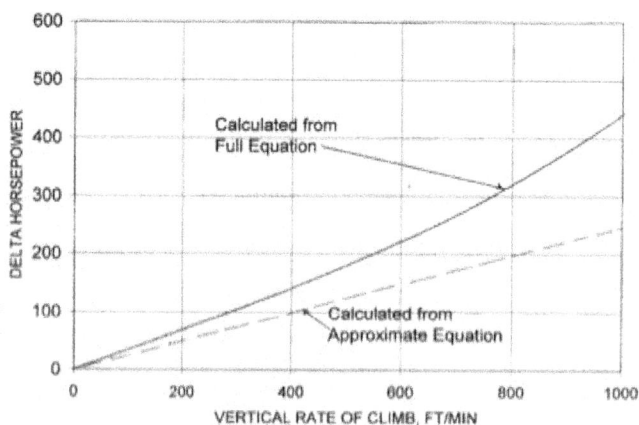

FIGURE 1. **Comparison of Results of Two Equations**

CHAPTER 4   *Trim and Control*

Any aircraft must obey the Equations of Equilibrium in steady flight with respect to its six Degrees of Freedom. How can we expect that a pilot will do the right thing if we give him only four controls?

Trimming the airplane

The explanation for the helicopter is more complicated than that for the airplane, so we will start with our "frozen wing "companion.

The six Degrees of Equilibrium have to do with the three possible forces in the direction of the axes shown in Figure 4-1 and the three possible moments about these axes. The system shown is referred to as the "Body Axis System "since it is locked into the aircraft's structure.

Airplane aerodynamicists usually use another system where the X-axis is pointed along the flight path as it would be in a wind tunnel, This is the "Wind Axis System. "Since we want the helicopter equations to work in hover where there is no "flight path "we use the Body Axis System. It is a simple job to go from one system to the other using trigonometry.

For an airplane in steady trimmed flight, all of the forces along each axis and all of the moments about each axis must be zero. Two of the six Degrees of Freedom are assumed satisfied by flying with zero sideslip and level wings which means that no deflections of the rudder or ailerons are required. (An "yes-but "to this is for high-powered airplanes with propellers which produce significant torque.) The other Degrees-of-Freedom are satisfied with correct settings of the throttle, and elevator. Maneuvering is done by using all these controls to produce unbalances in the forces or the moments to make the airplane do what the pilot wants.

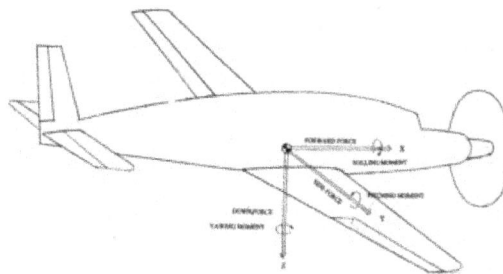

FIGURE 1. **The Airplane and its Axis System**

The symmetry of the airplane results in the fact that no deflections of the rudder or ailerons are needed for any normal steady flight condition, but one unique value of elevator setting will be required to trim out the inherent pitching moments.

Moving any control will initially produce an acceleration, but because of inherent damping, this will rapidly turn into a rate, or in the case of the elevator and rudder, a displacement with respect to the flight path. Pushing the throttle forward will produce more propeller thrust, but when the drag increases, a new speed "or rate "will be attained. Another rate control is associated with lateral stick. Holding the ailerons off their trim setting will produce a rate of roll as the wing acts as a windmill.

Elevator displacement is a mixed bag. A small amount of aft stick will pitch the airplane up, but the action will be stopped by the longitudinal stability provided by the horizontal stabilizer as if the airplane were restrained by a spring. The pilot, however, can use more aft stick to nullify this stability and do a loop.

Similarly, rudder deflection will produce an yawing acceleration, which changes into a yaw rate, and then into a sideslip angle as the directional stability produced by the vertical stabilizer stops the motion. Turns are done by banking the airplane, first by deflecting the ailerons to achieve a rate of roll, and then stopping it by neutralizing the control.

How about the Helicopter?

The helicopter is the same, but different. The same axis system as used for the airplane is shown in Figure 4-2. The pilot with his four cockpit controls has direct control of vertical forces with the collective stick, control of pitch and roll moments with cyclic pitch control as they produce blade flapping; and control of yawing moment with his "rudder pedals "

A significant difference between airplanes and single-rotor helicopters can be traced to the helicopter's lack of symmetry brought on by the presence of the tail rotor thrust. This requires, unlike the airplane in trimmed level flight, that some unique value of rudder pedal position and some unique value of lateral stick position are required for every steady flight condition.

Increasing a forward force is the result of a follow-on action involving pitching the helicopter nose down with longitudinal cyclic pitch and then stopping the action when the rotor is at the desired angle with respect to the horizon. An unique capability of the helicopter denied to the airplane is pure sideward flight. This is achieved by rolling the helicopter from hover and then stopping it at the desired rotor angle.

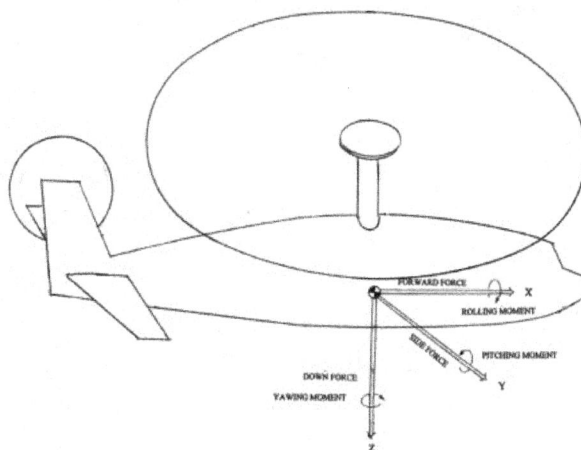

FORWARD FORCE — X
ROLLING MOMENT
SIDE FORCE — PITCHING MOMENT — Y
DOWN FORCE
YAWING MOMENT
Z

FIGURE 2. **The Helicopter and its Axis System**

Again, for maneuvering, changing control positions from trim values produces rates. This is especially true in hover. Longitudinal and lateral cyclic pitch produce pitch and roll rates as the unbalanced airloads cause the rotor to precess as a gyro at a rate that can be calculated from its parameters. Collective pitch changes the vertical rate of climb, and the tail rotor can change the rate of yaw in a hover turn.

In forward flight, moving the lateral cyclic from its trim value produces about the same rate of roll as in hover. If the helicopter has a large enough horizontal stabilizer, displacing the longitudinal stick a small amount, as on the airplane, will result in a pitch action that is stopped by the longitudinal stability. But by using more stick, the helicopter pilot can do a loop just as can his airplane pilot friend. Also, in flight at some low air speed, the helicopter pilot can turn his aircraft entirely around - a stunt I have never seen an airplane do.

Trim and Control

CHAPTER 5   *The Lift-to-Drag Ratio*

The airplane aerodynamicist makes good use of a parameter called the lift-to-drag ratio (L/D) for his aircraft. It is often used to compare one airplane to another, and has an important role in the Breguet Range Equation that tells him how far his airplane can go on a tank of fuel. It also tells him what the angle of descent is in a glide. When was the last time you saw a helicopter analysis based on L/D? Never?

The reason we don't find it useful is because helicopter aerodynamics are much more complicated than those of airplanes. For airplanes, the lift and drag coefficients are functions only of its angle of attack (neglecting stall and/or compressibility). At a certain angle of attack, the ratio of these two coefficients reaches a maximum. This angle can be found by testing a propeller-less model in a wind tunnel with results as shown in Figure 5-1. It can be shown that this optimum angle is where the parasite drag and the induced drag are equal. In flight, the condition is set up when the indicated airspeed that gives this angle is the one which also produces lift equal to the gross weight.

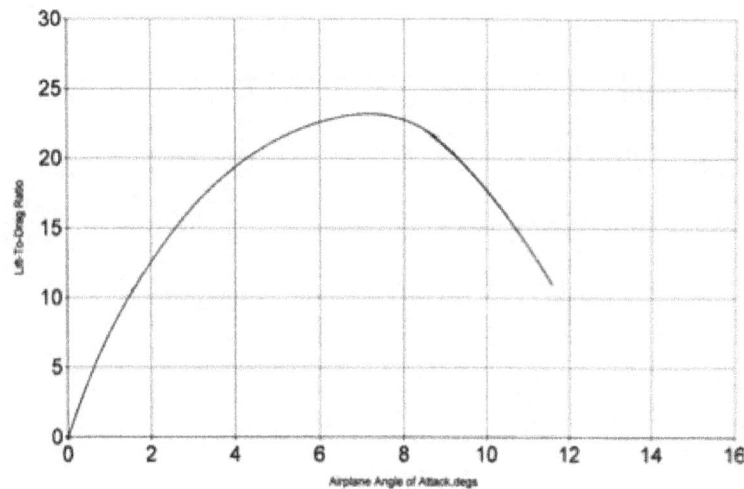

FIGURE 1. Airplane Lift-to-Drag Ratio

The angle of attack for maximum L/D for an airplane is fixed; i.e. it is independent of altitude and is the same in a climb or glide as in level flight. A way to determine the maximum airplane L/D in flight test is to look for the minimum glide angle as a function of indicated airspeed. On airplanes this will be penalized by the drag of a stopped propeller, but it works very nicely on sailplanes.

But we have it Harder

No such simple method exits for the helicopter. The rotor blades are working with many angles of attack and the rotor is also acting as a propeller. Finally, it is true airspeed, not indicated, that sets up the rotor aerodynamics in forward flight. Thus we have to work with the more complicated equations for power instead of those for drag.

To study the rotor alone, the helicopter aerodynamicist has a method of using his computer to calculating the lift-to-drag ratio of his rotor as if it were an airplane wing. This can be done by finding the total main rotor power required in forward flight and then subtracting out all the power that its propeller function is taking care of: those powers associated with parasite drag and the tail rotor H-Force. The remaining power is then converted to a drag value for plotting L/D

The results of calculations for my example helicopter are given in the Figure 5-2. Maximum calculated rotor values for L/D of six to eight are typical whereas for an equivalent wing, values higher than twenty can be achieved. This is an indication of why helicopters can't compete with airplanes going from Point A to Point B as long as those points have a mile or so of concrete for landings and take-offs.

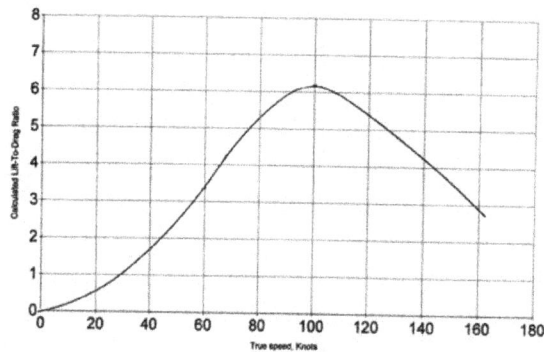

FIGURE 2. **Rotor Lift-to-Drag Ratio**

CHAPTER 6   *The Maximum Hover Figure of Merit*

The efficiency of hover performance is represented by the Figure of Merit which is the ratio of the induced power--which is associated with thrust--to the actual power which is the sum of the induced power plus the profile power, which is primarily due to blade skin friction.

Using special considerations, an equation for the maximum Figure of Merit can be derived as a function of only two parameters. The first represents the airfoil characteristics in the form of the ratio of the lift coefficient raised to the three-halves power divided by the drag coefficient, Cl 3/2/Cd. For simplicity, let's call this the "Efficiency Factor. "The Figure of Merit will be high when this parameter is high. Since the ratio is dependent on the angle of attack of a specific airfoil, there will be a certain angle of attack where it is a maximum.

This is shown on Figure 6-1 for two airfoils, the classic NACA 0012 used on many old designs and the more modern VR-7 used on the Boeing Chinook.

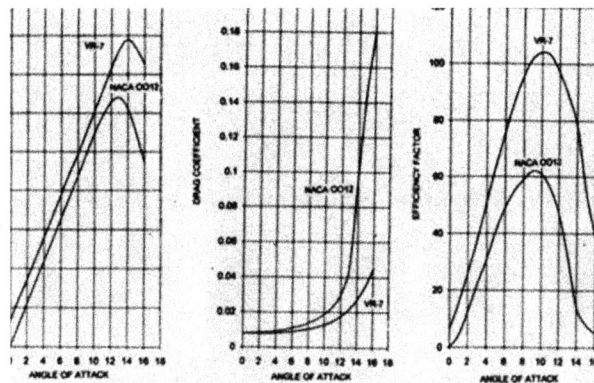

FIGURE 1. Airfoil Characteristic

It can be seen from the last part of Figure 6-1 that for both airfoils, an angle of attack of about ten degrees is optimum giving an Efficiency Factor of 62 for the NACA 0012 airfoil, and 104 for the VR-7.

The significance of this ratio is not limited to helicopters. If you were flying a sailplane and wanted to have the minimum rate of descent, you would fly at the airspeed where its value of CL 3/2/CD was the highest.

The second important parameter in the equation is the square root of the solidity ratio. The Figure of Merit will be high when this is high. By the way we design rotors the solidity is related in a general way to the disc loading, so rotors with high disc loadings will have high solidity and high Figures of Merits.

Two special considerations are used in the derivation of the equation. The first is that the induced velocity at each blade element is the same. This is achieved by "ideal twist "in which the pitch becomes steeper and steeper going from the tip to the root. The second is that each element is working at the angle of attack where the value of Efficiency Factor is a maximum, achieved with "ideal taper "where the chord becomes larger and larger going from the tip to the root. We are not likely to see such an odd-looking blades on real helicopters, but, hey, right now we are only looking for a theoretical maximum value to the Figure of Merit.

The result of evaluating the equation for the two airfoils is plotted as a function of rotor solidity on Figure 6-2. It shows that a Figure of Merit of 0.9 is a goal that we should consider, but no real rotor has as yet achieved anything close.

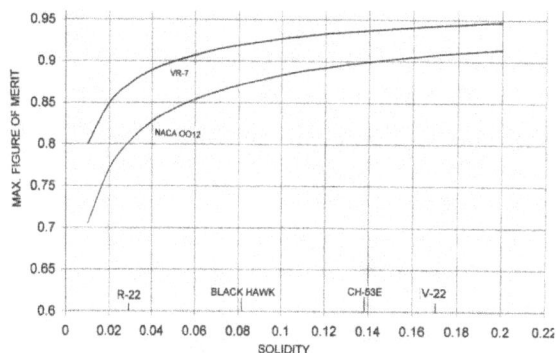

FIGURE 2. **The Theoretical Maximum Figure of Merit**

CHAPTER 7   *Who Carries the Load?*

For an airplane, each wing element does its good share in producing lift. This is true for a helicopter in hover, but not in forward flight.

Figure 7-1 shows the angle of attack and lift per running foot in hover for the blade elements of my example helicopter at sea level.

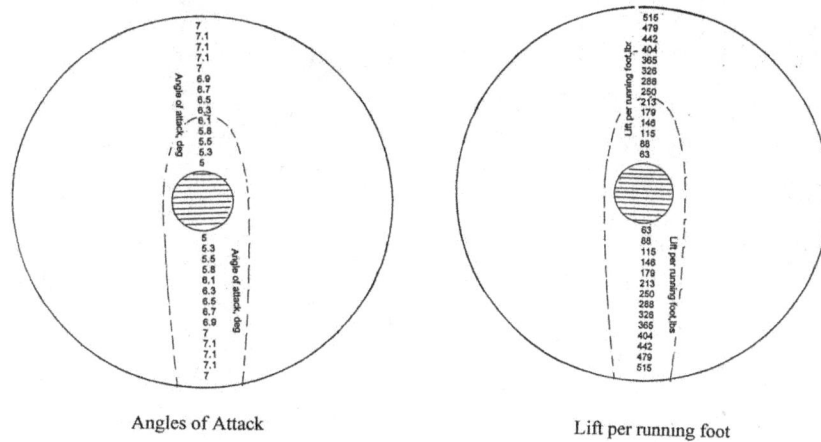

Angles of Attack   Lift per running foot

FIGURE 1. Conditions in Hover

The same parameters for forward flight at 115 knots are shown in Figure7-2. It can be seen that for the blades over the nose and tail, both parameters are nearly the same as in hover. In other words, forward speed has little effect on these blades: they think they are still in hover.

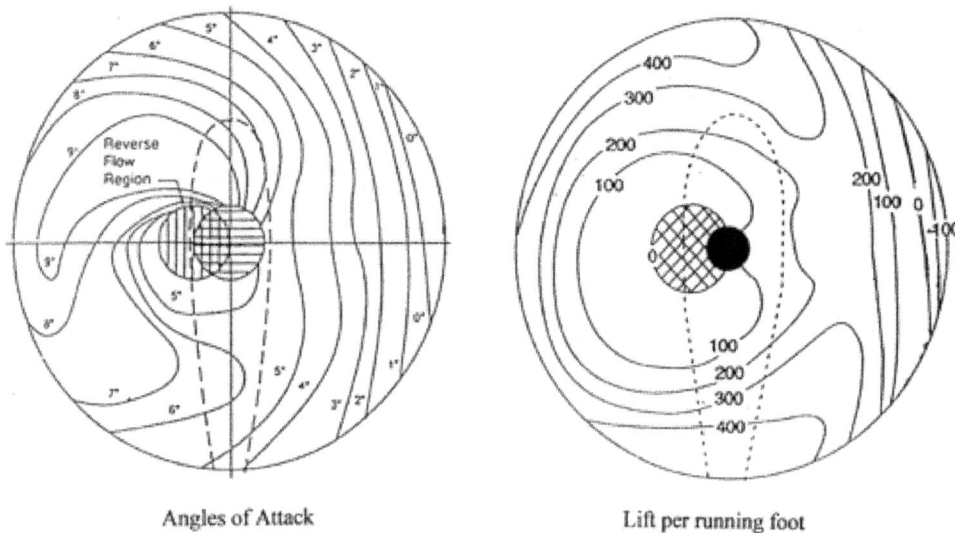

Angles of Attack   Lift per running foot

FIGURE 2. Conditions in Forward Flight

The blade elements on the retreating blade in forward flight cannot carry much lift even with high angles of attack because their dynamic pressures are so low. As a consequence, the advancing blade elements also can not carry much lift or the result would be an unwanted rolling moment. For this reason, the blades over the nose and tail get the assignment of doing most of the work.

For good rotor performance either in hover or forward flight, it would be desirable to fly each blade element at an angle of attack where its airfoil's lift-to-drag ratio was highest. For most airfoils this is around 10 degrees. The calculated angles of attack in hover for my helicopter are slightly less than this, but if it were to go to a higher altitude, they would approach the optimum.

At a forward speed at 115 knots, the angles of attack are variable; only locally approaching the optimum in the third quadrant. Flying faster or higher would make these angles approach stall. The situation looks more confusing on the advancing side where the angles at the tip actually go negative to produce a down-load--something you don't like to see on any aircraft.

This is the result of a law that applies to all helicopter design efforts. "Whatever helps hover hurts forward flight and vice versa. In the case of my example helicopter, the negative lift in forward flight is due to the desire to have good performance in hover. This led to making the blades with minus ten degrees of twist, and thus a resulting lift penalty at 115 knots.

(The other side of the coin is illustrated by using a retractable landing gear. The helicopter can fly faster, but at a weight penalty that reduces the permissible hover payload.)

The fact that not many of the blade elements are not operating near optimum angles of attack in forward flight is the reason that helicopters cannot compete with airplanes in going from Point A to Point B when each point has several thousand feet of concrete to operate from.

Another observation from Figure 7-2 can be made about the angle of attacks on the blades over the nose and tail. These blades carrying most of the weight have angles of attack that are about a third of what you would expect their stall angles would be. This indicates that during a pull-up maneuver, my helicopter is limited to a load factor of about three G's. This is about as high as I have ever seen from flight test with a helicopter flying at its design gross weight.

CHAPTER 8   *Ideal Twist*

Linear Twist

It has been well known for many years that the hovering performance of a helicopter can be improved by twisting the blades such that the pitch at the tip is lower than it is near the root. On most helicopters this "linear twist " scheme involves decreasing the pitch by a fixed amount per foot of blade going toward the tip. The value of twist the helicopter aerodynamicist uses is based on the difference between what would exist if the blade extended into the center of rotation and the pitch at the tip. On modern helicopters, it varies from -6 to -18 degrees

Ideal Twist

It has also been known for many years that there is another possible twist scheme. It is known as "ideal twist "and involves starting the pitch with a finite value at the tip and then dividing this value by the blade station ratio as you go inboard. It is thus nonlinear.

A comparison of the two twist schemes for the same rotor thrust is shown on Figure 8-1. Of course, if ideal twist were to be actually used on a blade, it would not be used inboard of about 25% of the radius where the pitch angle would be extreme as shown, but the main effect is happening on the outboard section 0f the blade.

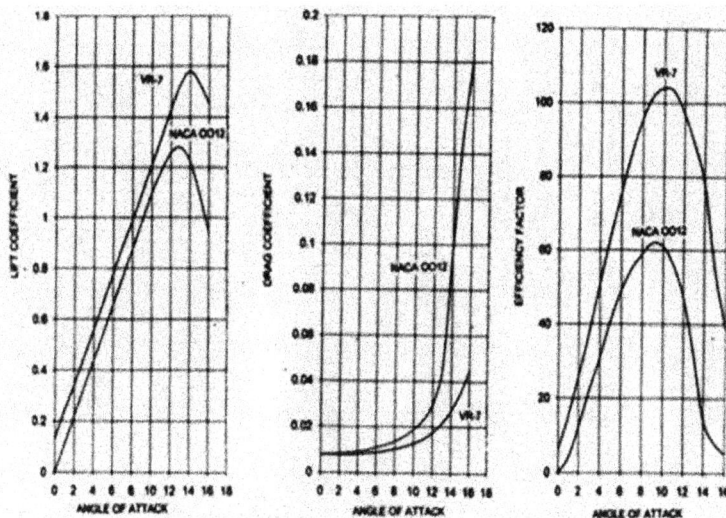

FIGURE 1. **Comparison of Twist Schemes**

Two Reasons for Considering Ideal Twist

When deriving the equations for rotor thrust in hover, the assumption of ideal twist simplifies the work by putting the term for twist in the same form as the angle of the downflow through the rotor disc.

Once ideal twist is used to write the equations for conditions at a blade element, it is easy to see that for a rotor with constant chord blades, the induced downflow all over the disc is uniform and the lift distribution is triangular rather than curved as it would be with linear twist.The single value for downwash over the disc of a hovering helicopter with ideal twist can be shown to correspond to the constant downwash behind an elliptical wing. In both cases, the result is a minimum required induced power.

Our New Choice

The use of linear twist was chosen when blades were being made of sheet metal since the sheets could be easily twisted in this way. Now that we are making composite blades in molds, we are free to use any twist distribution we want.

The Possible Result

I have used my example helicopter hovering at sea level to examine the effect of twist on the Figure of Merit. The results are shown in Figure 8-2. It may be seen that linear twist does increase the Figure of Merit, but that using ideal twist produces an even higher value. It is not much, but for blades made in a mold, it comes free!

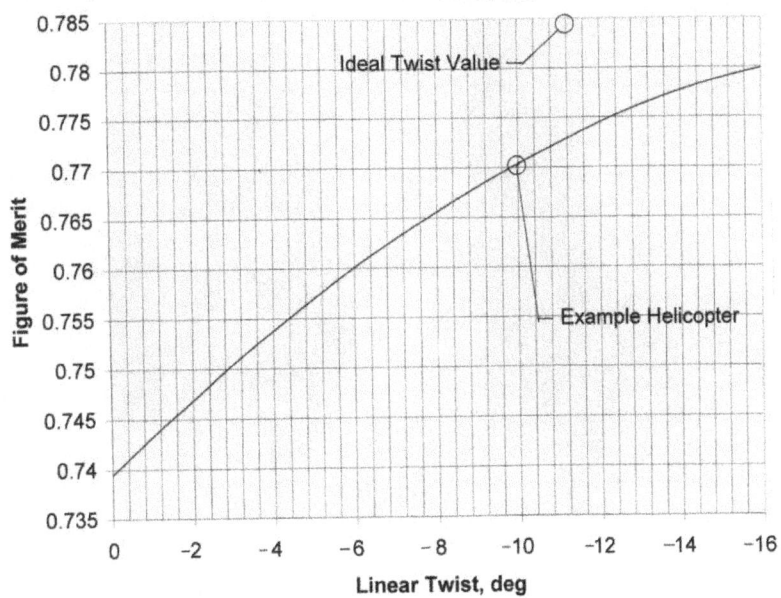

FIGURE 2. The Effect of Twist on Figure of Merit

Ideal Twist

CHAPTER 9   *Compressibility*

Everybody knows that air is compressible, but it has special significance to the aeronautical engineer. This is because an airplane or helicopter blade tips traveling at high speed through the air compresses the air ahead of it. As the aircraft flies through the air, it sends out compression waves to warn air molecules that something is coming their way and they should start moving to clear a path. The speed at which the compression waves advance is the speed of sound. The warning works well up to about 300 knots. At higher speeds, the molecules don't have time to react and we start getting "compression effects "

For high-speed airplanes, of course, this is a major consideration. Compressibility increases the drag on the entire aircraft. For helicopters the effects are localized to the blades, especially those on the advancing side where the tips may be operating close to the speed of sound to optimize performance.

The factor used in both fixed and rotary wing aerodynamics is the "Mach Number, "named after an Austrian physicist working in the early 1900s. It is the ratio of the speed through the air to the speed of sound. The speed of sound is a function of the air temperature. It is about 690 knots at 100o F and 590 knots at -40o F.

If the local speed of a blade tip element on the advancing side is near the speed of sound, the air going over the nose may be supersonic as shown in Figure 9-1 As the air goes over the nose of the airfoil, it speeds up and its Mach number increases. Further back, the air has to slow down. Air doesn't mind speeding up, but it doesn't like to slow down. If the local surface speed reaches a supersonic Mach number of about 1.4, the slowing-down process is almost instantaneously made through a shock wave. On airplanes, this produces high drag and also noise - the "sonic boom. "On helicopters, it produces the special "wop-wop "noise that tells you that a Bell UH-1 is approaching.

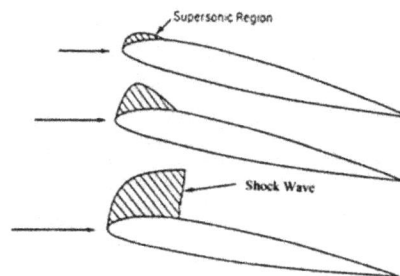

FIGURE 1. Airfoil with Local Supersonic Flow

35

The drag effect on a helicopter does cause the power required to increase, but the primary problem is the sudden nose-down pitching moment that is generated as the shock waves "both on the top and bottom surfaces "change their strength and location and thus the distribution of chordwise airloads on the blade. Helicopter aerodynamicists call this "Mach Tuck. "Helicopters with thick blades, such as the MDH 500 series with 15% thick blades, suffer more than helicopters with thinner blades because the air has to travel faster as it goes over the nose. Figure 9-2 shows Mach Tuck at zero lift for two airfoils as measured in a wind tunnel.

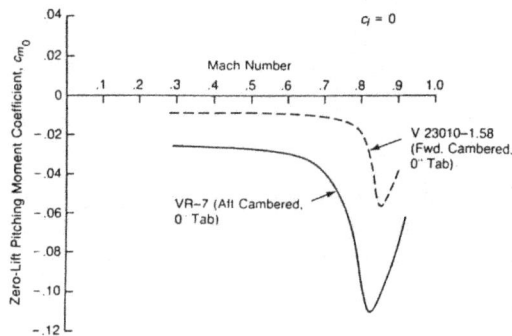

FIGURE 2. **Illustration of Mach Tuck**

The nose-down moment not only twists the blades, but must be reacted by the control system. The reason the Apache has swept blade tips is because with its original straight tips, the control forces at high speed were higher than the actuators had been designed for. The sweep fools the tip into thinking it is flying at a lower Mach number than it really is. This, of course, is the same effect taken advantage of on jet transports.

The Mach Tuck is only a minor consideration for airplane people, but because our blades are long and flexible, it becomes one of the limits for how fast a helicopter can fly. Blades tend to go out of track when shock waves are developed. Not every blade has the exact torsional stiffness as its mates. Thus when Mach Tuck comes, one will twist more than the others and generate a different spanwise loading and leave a different wake. That will have an effect on the others which will soon be flying on new paths of their own and the helicopter will experience a rough ride. The forward speed at which this happens is a function of the speed of sound and so will be more noticeable during winter flying in Alaska than over the Gulf of Mexico.

A wind tunnel test of a NACA 0012 airfoil at several transonic Mach Numbers shows how both the lift coefficient and the pitching moment coefficient are changed as a function of angle of attack. The cause is the changing shock wave strength and location on both the upper and lower surfaces.

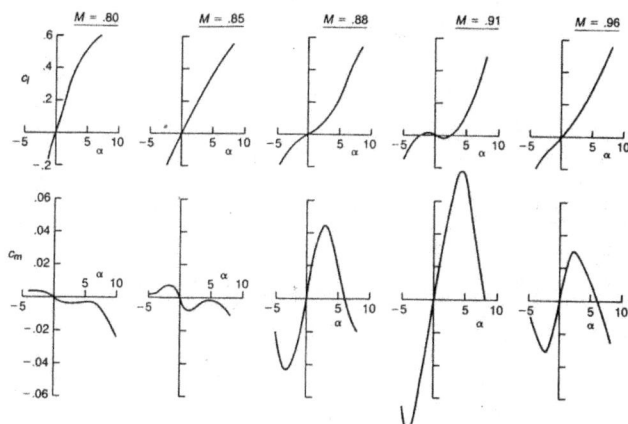

FIGURE 3. Lift and Pitching Moment Coefficients at High Mach Numbers

As can be seen, there is little effect at a Mach Number of 0.80, but at 0.91 there is a negative lift slope and a very high pitching moment which tend to disappear at 0.96.

The odd pitching moments twisted the blades of the Sikorsky compound helicopter to such an extent that they limited its top speed to 211 knots.

FIGURE 4. The Sikorsky Compound Helicopter

Compressibility

CHAPTER 10 *Another Look at the Advancing Blade Concept*

The Sikorsky announcement at the 2005 AHS Forum that the company will revisit the Advancing Blade Concept in the form of the X2 has made it timely to review the project that Sikorsky previously did with this configuration.

For those who don't know, the idea is to have a high angle of attack on the advancing blade and a low enough angle of attack on the retreating bade to eliminate retreating blade stall. On a conventional rotor this would produce an unacceptable unbalanced rolling moment, but if the rotor is very stiff, like a propeller, and is paired with another rotor doing the same thing but the other way, a total roll balance can be achieved. The result is a coaxial helicopter that can take advantage of the high dynamic pressure on the advancing side to produce high thrust.

In the 1970s, Sikorsky had the sponsorship of the U.S. military for its experimental aircraft designated the XH-59A or S-69. It first flew as a pure helicopter in 1973 and later with auxiliary propulsion from two jet engines as shown in the photograph. As a helicopter, it got up to a level flight speed of 160 knots and with the jets to 236 knots in level flight and 263 knots in a dive.

FIGURE 1. The Sikorsky ABC with Jet Engines

Letting it all Hang Out

A 1981 AHS Forum paper written by Arthur Linden and Andrew Ruddell at the end of the program lists the shortcomings of the XH-59A. These now become challenges for the designers of the X2.

High Loads

Of course one of the problems which might have been expected was high loads in each rotor and in the shafts between them as they fought each other by generating opposing rolling moments. Some planned flight test conditions had to be eliminated to avoid damaging stresses.

To hold the advancing tip Mach number to a value less than 0.95 at high forward speeds, it was proposed to slow the rotor down. This was found to be impossible because of the rapid rise of stresses as the tip speed ratio increased.

Besides the rolling moment, on the pure helicopter there were high stresses due to an aircraft pitching moment as speed increased since the horizontal stabilizer was mounted with positive incidence. It produced an ever-increasing nose-down moment that had to be trimmed out by the rotor producing a nose-up moment with corresponding high stresses. When the jet engines were added, the stabilizer incidence was set at a negative value to modify this characteristic.

High Drag

The authors reported that the rotor system had high drag. The two hubs and the shafts between them accounted for half of the total aircraft drag. (Note: the photograph shows the XH-59A with an instrumentation slip-ring assembly that made the situation even worse). Another source of drag was the inboard portions of the blades which had to be thick to provide the required blade stiffness.

High vibration

The XH-59A had high vibration levels. As the authors say: "they are well above those of typical production helicopters "

High Weight

The XH-59A was heavy. The ratio of empty weight to gross weight was 77.5%; a very high number when compared to modern production helicopters.

Control Problems

Like most other coaxial designs, yaw control was achieved using differential collective pitch. This is a system that works well in powered flight, but backwards in autorotation. The aircraft had large vertical stabilizing surfaces with rudders which were used for yaw control above 80 knots where the differential collective pitch system was phased out. Even with the rudders, auto-rotative tests showed a negative yaw control between 50 and 70 knots. (Why it was positive below 50 knots is not clear.) The authors blame part of the instability on the fact that since the surfaces were swept aft, they did not present a good aspect to the approaching air in autorotation.

Poor performance

Finally an unwelcome phenomenon that applies to all helicopter designs verified the saying that "Whatever helps hover hurts forward speed' and vice versa. In this case, at high forward speeds, the ten degrees of blade twist used to help hover performance produced a significant down-load on the outer part of the advancing blades. This, of course 'by the words in its title' is not what the configuration was supposed to do.

What about the X2?

The new helicopter configuration will be similar to the XH-59A except that the auxiliary propulsion will come from a pusher propeller instead of those fuel-hungry jet engines. The predicted cruise speed is 250 knots. Present Sikorsky engineers believe that they have the solutions to all those problems listed in 1981. Stay tuned to see if they do.

Another Look at the Advancing Blade Concept

CHAPTER 11   *What Should We Call the X2?*

When a botanist discovers a new plant and needs to name it, he looks at what makes it different from other plants of the same type--shape of the leaf or color of the flower and then looks in his Latin dictionary to find words that will tell another botanist what to look for.

In aviation, our pioneers did the same, but used Greek instead of Latin. They came up with aeroplane (air-surface), autogyro (self turning circle), and helicopter (helix-wing). For those with a knowledge of the Greek language, these names bring up a mental picture of the type of aircraft.

We now do the same in English--light plane, sailplane, jet fighter, jet transport, tilt rotor, and tilt wing among others.

The Russians do it too. For them a helicopter is a "vertolet "where "let "is Russian for wing. In German, "hubschrauber "means "heaving motion with a screw. "For people fluent in these languages, these names tell what type of aircraft is being talked about.

The Sikorsky X2 has gotten a lot of publicity lately, but this name does not bring up a mental picture. Even that of its predecessor, ABC, for Advancing Blade Concept does not help. Based on the table, I propose calling aircraft of the X2 type, "Semi Compound Helicopters "until somebody comes up with a better name.

Aircraft	Max Speed Year	Propulsion Unit	Wind	What We Call It
	216 Kts, 1986	No	No	Helicopter
	250 Kts, 2010	Yes	No	Semi-Compound Helicopter
	274 kts, 1969	Yes	Yes	Compound Helicopter

What Should We Call the X2?

CHAPTER 12   *The Cost of High Speed*

Sikorsky deserves congratulations for demonstrating a high speed flight of 250 knots with their "semi-compound "X2, but it should be realized that this desirable feature did not come cheaply The reason is embedded in the laws of physics. In this case it is the law that says that the power required to overcome the parasite drag of an aircraft is a function of the cube of the forward speed. For a given aircraft, if the speed requirement is doubled, the power plant has to provide the additional capability to compensate for the eight times increase in parasite power.

A Bit of History

The X2 is the revisit of the configuration flown by Sikorsky in the 1970s as the S-69 "Advancing Blade Concept "(ABC) aircraft. The idea behind this was that the maximum speeds of conventional helicopters were primarily limited by stall due to high angles of attack on the retreating side of the disc. To eliminate stall it seemed desirable to operate the retreating side at low angles of attack, but this would have produced a rolling moment. The S-69 avoided this by using two rigid counter-rotating rotors with low angles of attack on the retreating side and high on the advancing side. Each rotor produced a rolling moment, but since they are equal and opposite they compensated for each other.

The S-69 was not too successful. As originally flown as a "pure helicopter, "the maximum speed was only 160 knots. This was not the big improvement expected over other helicopters of the day. One reason was that because the blades were twisted to improve hover performance, the tips of the advancing blades at high speeds were actually producing negative lift rather than the positive lift that had been the basis of the idea.

This result can be traced to the requirement--as on all other helicopters--to use the rotor as a propeller as well as a wing. In later flying of the S-69, this requirement was eliminated by adding two jet engines. With these, the maximum level flight speed went to 236 knots. At normal helicopter speeds, the rotor of a conventional helicopter is working with enough air that it can easily do both the required lifting and propulsion jobs if provided with enough power. This capability exists at speeds up to about 200 knots when it simultaneously runs into stall problems on the retreating side and compressibility problems on the advancing side. (The current "pure helicopter "speed record set in 1986 is still 216 knots.)

For designing a rotary wing aircraft for speeds above 200 knots, it is necessary to provide some sort of propulsion device to relieve the rotor of the requirement to produce a forward thrust. This is undoubtedly the reason Sikorsky included the propeller in the X2 design (adding a wing could also help except in hover).

FIGURE 1. The Sikorsky X2

Choosing between a propeller and a jet engine is a subject of a trade-off. Jet engines are easy to install on existing helicopters and were used on the four compound helicopters flown in the 1960s during an Army program. But they do not work with much air and so must burn much fuel to produce the required thrust. The jet on the Lockheed compound XH-51--which got up to 263 knots--would suck the fuel tank dry in twenty minutes. A propeller working with much more air requires a significantly more modest fuel flow to do its job and this was the reason the Lockheed Cheyenne compound helicopter used a propeller instead of a jet engine.

A case study

We can illustrate the additional cost of high speed using the parameters of the Sikorsky X2 given in the table of page 20 of the Winter 2010 issue of Vertiflite.

The Hover Performance

First we will have to find the engine rating necessary to satisfy an Army requirement to hover as a 6100 pound pure coaxial helicopter at 6000 ft, 950 F at 95% of the Intermediate (30 minute) power available.

For this calculation, a program written for a conventional helicopter has been modified for this coaxial helicopter by eliminating the penalty of the tail rotor and of the rotation of the rotor wake. The calculated result is that 807 horsepower is required to satisfy the hover requirement. To obtain the sea level Intermediate rating, we add 5% to account for the specified 95% requirement and then add 32% of that result to account for how much a typical turboshaft engine's power rating decreases going from sea level on a standard day up to 6000 ft. on a 95 0 F day.

The answer is that an engine with a sea level Intermediate rating of 850 horsepower is needed to satisfy the hover requirement. The take-off rating (5 minutes) would be 5% more or about 900 horsepower. This means that the 1630 horsepower T800-LHT-801 engine installed the X-2 should have no trouble meeting the hover requirement.

What about Forward Speed?

The next step is to determine what the maximum forward flight speed difference would be for the X2 with and without a propeller. An important parameter for this analysis is the equivalent flat plate area which governs the magnitude of the parasite power.

A forward flight program for a compound helicopter--like the Lockheed Cheyenne--has been modified to represent the X2 configuration by eliminating the wing and the tail rotor. For this calculation, it has been assumed that the propeller efficiency is 85% and that the rotor tip speed is 525 feet per second to keep the Mach number at the advancing tip less than 0.9 at 250 knots. The fact that the program assumes a single main rotor while the X2 has a coaxial rotor is assumed to make no significant difference for this task. On a standard day, a typical turboshaft engine's take-off rating at 10,000 feet decreases to 83% of its sea level rating, or in this case to 1350 horse-power.

Running the program with these assumptions and with several values of equivalent flat plate area leads to the conclusion that 4 square feet is the value that requires 1350 horsepower at 250 knots at 10,000 feet. This seems to be a reasonable value considering the very streamlined X2 fuselage. Most of the parasitic drag is undoubtedly generated by the two rotor hubs. (For a point of reference, the equivalent flat plate area of the Hughes OH6-A as determined by full-scale wind tunnel tests is 5 square feet.)

The next step is to determine how fast the pure helicopter version of the X2 could go at 10,000 feet using an engine with a sea level takeoff rating of 900 horsepower (reduced to 750 at altitude) and an equivalent flat plate area of 4 square feet. This was done with the pure helicopter version of the compound helicopter program. The result of the calculation is a speed of 180 knots.

Thus the conclusion is that in the case of an aircraft like the Sikorsky X2, raising the maximum speed from 180 to 250 knots requires installing a propeller and raising the sea level take-off rating from 900 to 1630 horsepower.

Besides the cost of these two decisions, there is another factor that will increase the cost of operation over that of a slower pure helicopter. For hover and all other flight conditions except high speed, the engine will be working at partial power. To keep its big compressor running requires about the same amount of fuel flow no matter how much power is being generated. So if a mission includes significant time at low speeds, the fact that the helicopter has been designed for high speed will require extra fuel.

A possible way of getting around this situation is to use two 815 horsepower engines and to increase the rotor diameter a little so that the hover requirement would be met with one engine and then using both only when high speed was required.

And in conclusion

It may be seen that because of the additional cost, there must be a strong motivation to design a helicopter to fly faster than the current ones.

CHAPTER 13   *Designing for Balance*

An article in the July, 2012 issue of Rotor &Wing magazine specified three goals for future Army helicopters:

1. A maximum speed of more than 250 knots

2. A minimum hover power loading of 11 pounds per horsepower

3. A 20% improvement in empty weight fraction

How do We Do This?

A well-designed vertical take-off aircraft must meet its specified requirements both in hover and for maximum forward speed. We call this "balanced performance. "This can be illustrated using the classical plot of power loading (pounds of rotor thrust per horsepower) versus disc loading (pounds per square foot of disc area) developed from hover momentum physics as shown in the figure with three helicopters (each arbitrarily plotted at a Figure of Merit of 0.6). It can be seen that high power loading goes with low disc loading.

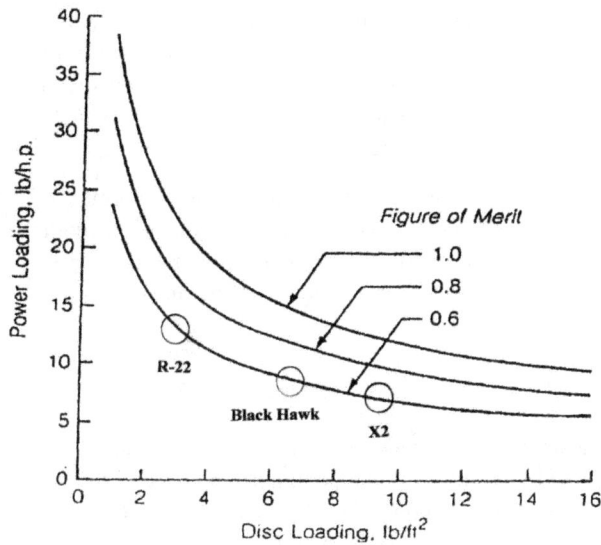

FIGURE 1. Hover Parameters

The R-22

The Robinson R-22 doesn't need a large engine to reach its cruise speed of 96 knots with 75% power so it can use a small engine. To get good hover performance from that engine, however, the rotor must be relatively large to produce a low disc loading of 3 in order to get a large amount of rotor thrust from each horsepower--12 pounds per horsepower for the R-22.

An Aside

For all aircraft, the lifting performance is governed by how much air is affected in flight. The more air, the less the rotor, propeller, or wing has to work on it and thus the less power is required. The airplane aerodynamicist will use a long wing to take advantage of "low-span loading. "This is evident on jet transports and especially on high-performance sailplanes. The challenge here is to design a long wing that will not break off during maneuvers or flying through gusty air. We rotary-wing people avoid this by using flapping hinges and the effects of centrifugal forces.

The Black Hawk

The next helicopter on the plot is the Sikorsky Black Hawk which cruises at a speed of 160 knots. To meet the Army goal of 250 knots, the engine rating must be increased. When high speed is a primary goal, the first big design decision is the size of the engine(s). The power required is the sum of the induced, profile, parasite, and tail rotor powers. At high speed, the parasite power is the largest component. Parasite power increases as the cube of the forward speed, so going from 160 knots to 250 increases the parasite power by a factor of almost 4. This will require an engine rating of about three times more than on a 160 knot aircraft.

Once a big enough engine has been selected to fly fast, getting adequate hover performance is not a problem. It is done by simply determining the rotor diameter at which the hover require-ment can be met. For a 250 knot Black Hawk, this will result in a smaller rotor and increased disc loading with a much smaller power loading than the present 8 pounds per horsepower and much less than the Army goal of 11.

A Possible Modification

The 160 knot Black Hawk could be redesigned to get the Army's desired power loading of 11 pounds per horsepower by decreasing the disc loading to 3 which requires increasing the rotor diameter from 53 feet 8 inches to 85 feet, but this would not be a balanced design. The hover ceiling with those two T-700 engines would be half-way to the moon.

Another Aside

During the life of a successful helicopter, the three factors that affect where it plots on the fig-ure: mission gross weight, rotor diameter, and engine power rating will probably change. For example, during the Black Hawk's early preliminary design phase, the mission gross weight was assumed to be 15,500 pounds, but the final contract was for 16,450, and the prototype, YUH-60A--after solving problems revealed by flight test--weighed 16,750. Finally, at the last delivery of the UH-60A in 1989, the mission design weight had climbed to 16,803 pounds. This 1300 pound growth is not unexpected. (An extreme example was the Comanche which, in twenty years, grew from 8,500 pounds to 12,349 without reaching production.)

During the development and delivery time of the UH-60A, the 53-foot diameter rotor was twice expanded by four inches, and the T-700's 30-minute engine rating had gone from 1622 to 1890 horsepower.

The X2

The Sikorsky X2 with a top speed of 250 knots and a gross weight of 6100 pounds and a 1630 horsepower engine has a power loading of 3.7. Getting the power loading up to 11 pounds per horsepower could be done as on the Black Hawk.

Empty Weight Considerations

The desire to reduce the empty weight fraction by 20% on a fast helicopter is a real challenge. Compared to a slower aircraft, any one designed for over 200 knots will require a propulsion system in the form of a propeller, ducted fan, or a super by-pass jet engine. Some designs will also have wings. The drive system will have to be big enough to handle the increased engine power with the corresponding weight. Retracting the landing gear will require a weighty mechanism. Another weight item is a stronger windshield since going from 160 to 250 knots more than doubles the dynamic pressure. Side windows must also be heavier; a side-slip at 250 knots will produce large suction forces on one side. If weapons are carried internally and only exposed when they are needed, the required mechanisms will be a weight penalty. And as a final weight increase, the bigger engines will be burning much fuel so the fuel tanks must be big to do the same missions that the slower helicopters do.

But the 250 knot helicopter will do the missions faster. For example, a 300 nautical mile mission done at 250 knots takes only an hour and 12 minutes instead of an hour and 50 minutes at 160 knots. Is that the over-riding consideration?

Designing for Balance

CHAPTER 14   *Evolution of a Compound Helicopter*

With the recent demonstration of high speeds by the Sikorsky X2 and the Eurocopter X3, there is an anticipation that the next generation of rotary-wing aircraft will be faster than the ones we know today.

FIGURE 1.

To illustrate some of the challenges involved, I have thought about how to make the conventional example helicopter of my textbook into a compound helicopter, raising the top speed from 185 to 285 knots at an altitude of 5000 feet at the engine's 30-minute rating.

Making Changes

This would be done by adding a wing and a pusher propeller to make a configuration like the Lockheed Cheyenne of the 1960's. (Making an advanced configuration from an existing design is the modern way. No new helicopter design has reached the production stage for many years.) I am going to retain the main and tail rotors of my conventional helicopter. Since the new one will undoubtedly have some low-speed mission segments, I will even keep the blade twist that helps hover performance.

FIGURE 2. Lockheed Cheyenne

When fully loaded, my conventional helicopter has a gross weight of 20,000 pounds and carries 30 passengers at a cruise speed of 172 knots. As a high-speed compound helicopter at this same gross weight it would also carry passengers, "but not so many. This is because of the increase in empty weight due to the addition of the wing, propeller, landing gear retraction mechanisms, bigger engines, and stronger drive systems. Another requirement that would add weight is stronger windshields and side windows to be able to survive with more than twice the dynamic pressure. The increased fuel to do a same mission at the higher cruise speed of 270 knots would also reduce the payload.

Even with a decrease in the number of passengers, the fuselage would have about the same shape and basic drag, but drag would be reduced by retracting the landing gear and using a thin door-hinge main rotor hub as on the Cheyenne. The change to the landing gear should give a reduction of 6.5 square feet of equivalent flat plate area and the hub redesign an additional 2.5 square feet resulting in reducing the equivalent flat plate area from 19.3 square feet to 10.3.

The Wing and Propeller

My criterion for the wing is that it should be able to support the full gross weight at 160 knots at sea level at a lift coefficient of 1.0. This requires 117 square feet. With an aspect ratio of 10, the wing span is 34 feet, just over half the rotor diameter of 60 feet. Of course, the purpose of the wing is to relieve the lift requirement on the rotor while the propeller relieves it of the requirement to provide a forward propulsive force. Each of these reduces the possibility of retreating blade stall.

A propeller with a diameter of 10 feet should be adequate. It will be assumed for this study that for all speeds, its efficiency is 80%.

Changing Rotor Thrust

At high tip speed ratios, the rotor is subject to retreating blade stall so it has to be unloaded. I propose to do this by reducing its collective pitch. For this early in the project, I have selected a reduction, starting at 100 knots, of a tenth of a degree per knot until the average pitch of the blade element at the 75% radius position is zero. In this condition, any change in rotor thrust is developed by a positive tip path plane angle of attack like an autogyro. At maximum speed, the rotor is calculated to be carrying about 20% of the gross weight with reasonable efficiency since it operates with a large amount of air.

Changing Tip Speed

Another goal is to avoid compressibility effects on the advancing blade tip. This is done by limiting the Mach Number at the tip to 0.85 by reducing the tip speed from 650 feet per second to 470 starting at about 180 knots. At 285 knots the tip speed ratio would be 1.02.

There would be a challenge in reducing the rotor RPM by 30%. We must find out from the Dynamic Engineers where the blade natural frequencies are to avoid dwelling on resonance points as we reduce rotor speed. Should the reduction be done with the engine governors as on the Osprey or with a gear shift as on the Bell X-3? Should the propeller be driven from the main rotor transmission or from a separate gear box with a variable gear ratio? These decisions are to be made later.

The Bottom Line

The calculations show that at 285 knots at 5000 ft, this aircraft needs engine power of 4000 horsepower which despite the reduction in flat plate area, is higher than the 3400 needed in the conventional 185 knot helicopter. The power of these bigger engines would probably be just enough to compensate for the aerodynamic download of the wing in hover.

These are "first day "decisions just to get started. They would certainly be modified as the project develops.

Evolution of a Compound Helicopter

CHAPTER 15   *Sizing the Osprey*

One of the first parameters needed when starting out on new helicopter design is the disc loading. On conventional helicopters the procedure is fairly straight-forward. The requirements would be known: including the payload, the mission, the maximum speed and the hover performance. Based on a knowledge of existing helicopters, the payload and mission can be used to estimate the gross weight and drag characteristics. The next step is to estimate how much power is needed to meet the high speed requirement. With the engine(s) that does this, what is the smallest rotor diameter that satisfies the hover performance? With the estimated gross weight, this sets the first estimate for disc loading to give a minimum-sized "balanced "design that just barely meets both the high speed and hover requirements.

For the balanced Black Hawk and Apache configurations with maximum speeds of about 150 knots, the disc loading is about nine pounds per square foot, but for the CH-53E, with three big engines to make 170 knots, hover is no problem and so the resulting disc loading is higher, at 14.

A New Requirement

It would seem that the same procedure could have been used on the V-22 Osprey, but it wasn't. A special requirement for this aircraft was that it had to be compatible with the small carriers that it would be operating from. A critical "flight "condition was taxiing the length of the flight deck without running into the island or falling overboard. Shipboard compatibility tests with conventional helicopters were made and safety considerations resulted in the requirement that the rotor tips had to be at least fifteen feet from the island and the outboard wheel could be no closer than five feet from the outer edge of the deck.

FIGURE 1.

For the smallest carrier that the Marines were considering as an Osprey operating base, this established a maximum rotor "span "of 85 feet. The fuselage width is established by the payload and the distance between the rotor tips is determined by the clearance from the fuselage in airplane flight. Thus the rotor diameters were limited to 38 feet. At a gross weight of 46,000 pounds, the resulting disc loading is 20 pounds per square foot.

This is enough to generate hurricane-like velocities in the rotor wake with resulting problems when operating close to any ground softer than concrete.

Lesson Learned

This is just an example of how special requirements other than performance can influence the design. The Bell HSL anti-submarine tandem-rotor helicopter, unlike other tandems, had its two rotors at the same height so that it could be stowed in the limited ceiling height of the carrier's on-board hanger. The result was high noise coming from blade-vortex interference even in normal hover and forward flight conditions.

CHAPTER 16   *Canted Tail Rotors*

How It All Began

The first canted tail rotor appeared on the Sikorsky Black Hawk as a result of having to meet an Army requirement. When writing the specifications for the Utility Tactical Transport System (UTTAS) competition, a requirement was included that one of these helicopters must fit into an Air Force C-130 cargo airplane and two should fit into a C-141. The preliminary design process for the Black Hawk had been completed. The diameters of the main and tail rotors had been chosen to give the performance required. That set the tail boom length and the length of the nose had been designed to put the center of gravity close to the rotor shaft. As a last check, two small cardboard planforms of the design were placed on the floor plan of the C-141's cargo compartment, and they did not fit!

Back to the Drawing Board

The solution was to shorten the nose. This put the center of gravity well behind the main rotor mast. For airplanes, a center of gravity behind the center of lift can be a serious problem by leading to an aircraft that is longitudinally unstable. The situation is different for helicopters because of the way that rotor flapping is used to trim out the pitching moments about the center of gravity. An aft C.G. does not cause instability. (For a discussion of this, see my column in the March 1997 issue of Rotor and Wing, or Chapter 48 of my book, Helicopter Aerodynamics, Volume II.)

So the aft C.G. on the Black Hawk is not a stability issue, but it is bad in that the nose-down flapping to balance moments about the center of gravity produces high fatigue loads in the rotor hub. The Sikorsky solution was to cant the tail rotor by twenty degrees so that its upward force would produce a nose-down moment about the center of gravity opposite to the nose-up moment from the main rotor thrust. This reduced the flapping required for trim.

FIGURE 1. The Sikorsky Black Hawk

Which Side?

American aerodynamicists would like to see the tail rotor mounted on the left side of the fin. In this "pusher "arrangement, it is sucking air past the fin instead of blowing on it as it would be if it were mounted on the right side of the fin as a "tractor. "As a tractor, the opposing fin force makes the tail rotor work harder to do the anti-torque job than as a pusher and the additional power required is significant. Despite this, Sikorsky chose to use the tractor position to assure more clearance with the tail boom if something went wrong.

Couplings

Another result of this design choice is a coupling that makes the helicopter pitch as a result of moving the rudder pedals. A push on the left pedal to start a left turn will increase tail rotor thrust, and its upward component will pitch the helicopter nose-down. Another coupling is due to the main rotor being ahead of the center of gravity. An increase in collective pitch will produce a nose-up pitching moment. These couplings are reduced by using a rather complicated and heavy mixing box in the control system.

Yet another coupling due to the canted tail rotor is produced by sideslip. In an inadvertent slideslip to the right, the tail rotor thrust will be reduced along with its vertical component. This will result in a nose-up pitching moment about the center of gravity. Sikorsky chose not to correct this with the rotor, but to use the stabilator instead. An accelerometer in its computer senses lateral acceleration and changes the incidence--positive in the case of an inadvertent right sideslip.

I once heard a Sikorsky engineer giving a paper on the Black Hawk control system and these complications that were introduced by the canted tail rotor. He concluded his verbal presentation by saying, "We'll never do that again. "The response to that opinion these days is "Well, all modern helicopters will be fly-by-wire "

Sikorsky later used the canted tail rotor on the CH-53E which was also tail-heavy. A difference was that it had a canted fin so that the tail rotor could be mounted on the left side with plenty of clearance.

FIGURE 2. The Sikorsky CH-53E

The Comanche also had a canted ducted fan, but the motivation for this aircraft was to decrease its radar signature.

FIGURE 3. **The Comanche**

Other benefits

The shortening of the nose on the Black Hawk saved a little structural weight and reduced the aerodynamic download in hover.

There is yet another significant effect on hover performance. This is the result of trigonometry functions. The thrust on the canted tail rotor must be higher than without cant by 1/cosine of the cant angle. For twenty degrees, this factor is 1.06. This accounts for a modest increase in required tail rotor power to do the anti-torque job. On the other hand, the vertical component of thrust is proportional to the sine of the cant angle, or 34%. This is the amount of tail rotor thrust that can be used to relieve the thrust and power requirements of the main rotor.

I have calculated the hover performance of my example helicopter at its design gross weight of 20,000 pounds at sea level with and without a 20 degree cant angle. With the cant, the main rotor power required is 59 horsepower less than without cant, but the tail rotor power is 8 horsepower more. Since the power loading is 9 pounds per horsepower, I could have increased the payload of my design by about 450 pounds (two passengers) by using tail rotor cant.

Yes, but...

When I used my forward flight program on my example helicopter, I got another result. The power required was higher with the 20 degrees of tail rotor cant than without it. This is just another example of the designer's dilemma: "Whatever helps hover, hurts forward flight "

Whereas in hover the tail rotor lift was beneficial by helping the main rotor hold the aircraft up, in forward flight, its effect on trimming the pitching moments about the center of gravity was not beneficial. This is because the lift of the tail rotor made a nose-down pitching moment which must be balanced by nose-up rotor flapping. But the attitude of the rotor with respect to the horizon is essentially the same, so the fuselage will fly more nose-down and thus have higher drag than without cant.

For my helicopter at 150 knots, the power difference is 540 horsepower. For a three hour mission, this requires an additional 750 pounds of fuel. Oh-oh! there goes my gain in payload.

A possible help for this situation is to carry more download on the horizontal stabilizer.

A New Trend?

There are three other designs now using tail rotor cant. The Army is studying Future Multi-Role Helicopters. This design uses a canted tail rotor to take advantage of the increase in hover performance.

FIGURE 4. **An Army Study for a Future Vertical Lift Helicopter**

The other two are Bell's Model 525, and Eurocopter's EC 175.

FIGURE 5. **The Bell 525 The EC 175**

Because the EC 175's main rotor is turning in the "French "direction, the cant is in the opposite direction as would be used on American helicopters.

Whether this design feature will appear on a new helicopter seems to depend on its critical operational requirement: hover or cruise?

CHAPTER 17   *Variable Speed Rotors*

The rotor speed on most helicopters is meant to be kept constant 'or nearly so "for a very good reason which will be discussed at the end of this chapter. But what if that reason did not exist? Would we consider varying the rotor speed in flight to get better performance? You bet we would!

Considering that we have that option, here are four flight conditions that might benefit:

1. Hover out of ground effect
2. Loiter at speed for maximum endurance
3. Cruise at speed for maximum range
4. Maximum speed at full power

The objective of using a variable rotor speed is to make as many blade elements as possible operate at or near the angle of attack for the maximum lift' to-drag ratio of the airfoil. This is in the region of six to ten degrees for airfoils used on rotor blades. A reflection of the average angle of attack over the rotor disc is found in the blade loading coefficient, C_T/σ, with a value of 0.13 representing about eight degrees. The designer has both the rotor speed and the blade area as parameters to adjust in order to achieve that goal, but in this study only the rotor speed has been used as a variable.

The calculations have been based on the twin turbine-powered example helicopter of my text book. The only change has been to use the VR-7 airfoil (as on the Chinook) instead of the NACA 0012 to get better high-speed performance by delaying retreating blade stall. The helicopter has a tip speed of 650 ft/sec which will be referred to as 100% RPM. All calculations have been done at sea level and at altitudes of 10,000 and 20,000 feet.

Hover

The index of "goodness "in hover is the Figure of Merit, which is like an efficiency rating. The hover program computes the Figure of Merit at the same rotor thrust as a function of percent RPM. Figure 17-1 shows the results for the three altitudes. It may be seen that even though the Figure of Merit changes with altitude, the blade loading coefficient at each maximum value is about the same. At sea level, it would be desirable to lower the rotor speed to 82% to achieve the maximum hover performance, but at altitude, the 100% RPM value is about as good as it gets.

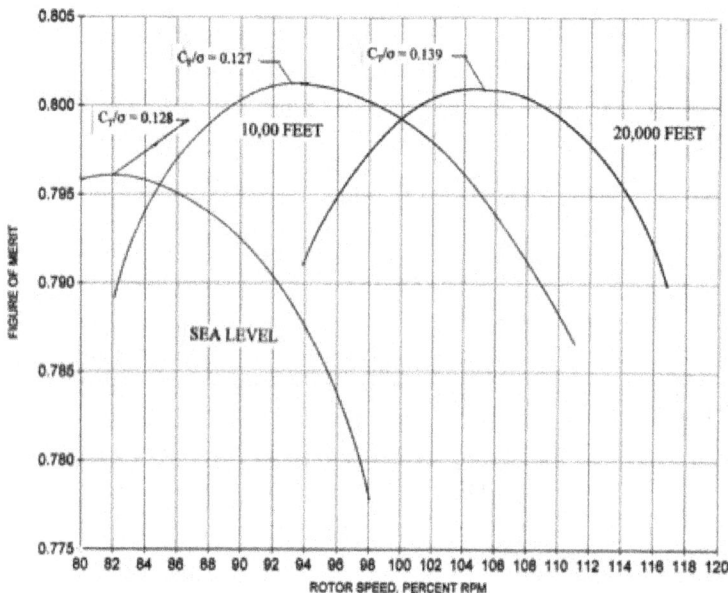

GURE 1. Hover Figure of Merit

Loiter

For this flight condition, the pertinent measurement is the Specific Endurance (S.E.) measured as hours of flight per pound of fuel. For mission planning, this tells the operator the minimum fuel to load for a required "time on station. "To do this analysis, we must know the characteristics of the engine. I have assumed that the fuel flow as a function of power and altitude is as shown on Figure 17-2. Note that the intercepts at zero power depend on altitude. The turbine engine benefits from operation at high altitudes because with thinner air the compressor requires less power. (And this is one of the reasons that jet transports fly so high.)

The maximum S.E. will be achieved at the forward speed for minimum power - the "bucket speed. "The program computes the power required at a given altitude, percent RPM, and at a series of forward speeds. It then finds the fuel flow in pounds per hour that goes with those flight conditions and takes the inverse of that number to arrive at the S.E. in hours per pound of fuel. For Figure 17-3, the maximum value is plotted with a notation about the forward speed and blade loading coefficient. The optimum values of C_T/s for maximum endurance are only slightly lower than the optimum values for Figure of Merit indicating that at these low speeds, the rotor aerodynamics have changed only a little

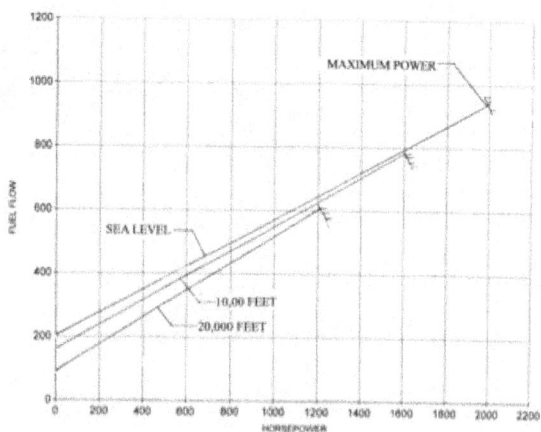

FIGURE 2. Assumed Fuel Flow Characteristics (lb/hr)

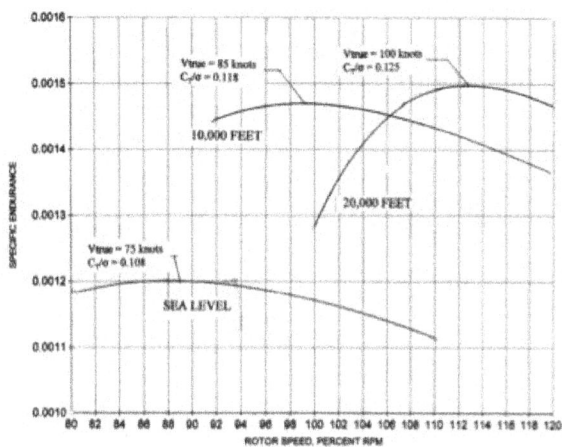

FIGURE 3. Specific Endurance (hr/lb)

Cruise

The factor of importance for cruise is the Specific Range (S.R.) which is given as nautical miles per pound of fuel. Knowing this, the minimum fuel load for a required range can be estimated. The computation is essentially the same as for loiter except that the S.R. is equal to the forward speed divided by the fuel flow to give nautical miles per pound of fuel. Maximum values are obtained at speeds considerably higher than for loiter.

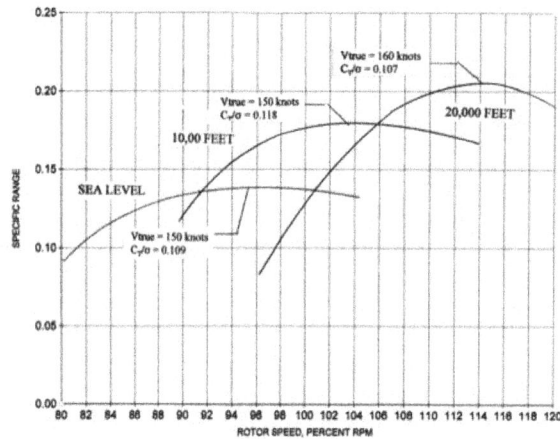

FIGURE 4. Specific Range (n.mi/lb)

The calculations for Figure 17-4 were made at the same time as those for the specific endurance of Figure 17-3. At these higher speeds, the optimum values of the blade loading coefficient are lower than for the prior two conditions indicating that some blade elements on the retreating side are operating at angles of attack higher than optimum and most on the advancing side are operating at lower angles than optimum.

Maximum Speed

Maximum speed is either when the power required is equal to the power available or when retreating blade stall makes it impossible to trim the rotor with cyclic pitch. Figure 17-5 shows both limitations as a function of rotor speed. For this condition, the blade loading coefficient is lower that any of the other conditions indicating some degree of retreating blade stall at each altitude.

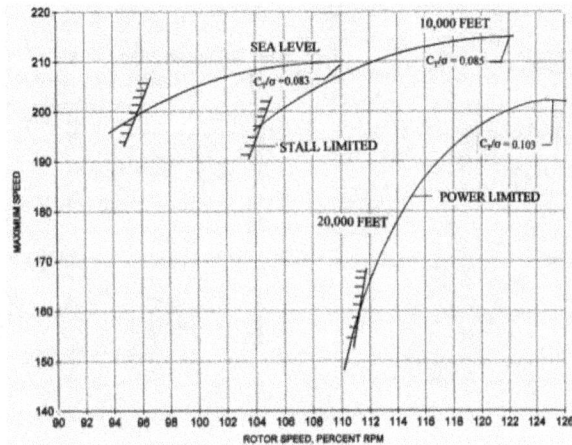

FIGURE 5. Maximum Speed

Why we don't use Variable Rotor Speeds

All helicopters shake! They shake even harder if some component is being excited at its natural frequency, a condition known as "resonance. "Structural components including everything from blades to the entire fuselage to the side windows, like the strings on a violin, have their own individual natural frequencies that may be in step with the passage of the main or tail rotors blades or multiples of these passage frequencies. These conditions are found on all helicopters during initial flight tests. They can vary from annoying if they only produce high vibration to dangerous if they produce high stresses. They provide a challenge to the dynamics engineer who must try to adjust the natural frequencies so that they are not in tune with the rotor speed that had been chosen as 100%. If he is successful at this speed, he will probably be frustrated if that speed is changed.

The element most likely to cause trouble is the rotor blade itself since the natural frequencies of its bending "modes "depends on centrifugal forces that change with rotor speed. An illustration of this phenomenon is given in Figure 17-6 by showing how a typical blade can bend and how its natural frequencies of flapping, and bending in various directions can be plotted as solid lines against rotor speed. The dashed lines represent multiples of rotational frequency from once per revolution (1P) to 6P for this plot. Their pattern inspires calling this a "fan plot, "but its proper name is a "Southwell Diagram' after the dynamicist who first used it. In forward flight, there are aerodynamic inputs that repeat themselves along each of these dashed lines. Where the solid lines are at or near a dashed line, a resonant condition exists and the blade response can be very high. This can lead to both high vibration and high blade stresses.

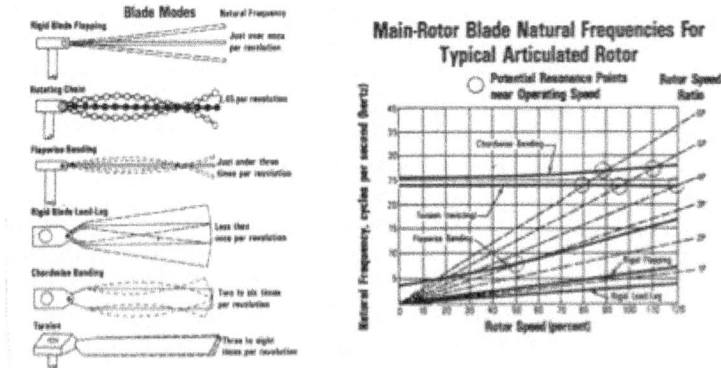

FIGURE 6. **Blade Modes and Natural Frequencies**

The blade shown is in fairly good shape at 100% RPM (though103% would be better), but could get rough at 79, 89, 96,110, and 119%. This leaves only four narrow regions where you might be able to run this rotor without major problems, and you would want to quickly pass through any of the resonant speeds while going from one gap to the other. (The resonance condition at 50% is why some helicopters shake on windy days while passing through this rotor speed during the spin-up.)

The above shows why we are reluctant to use much variation in rotor speed. It does not, however, say that it is obviously impossible.

CHAPTER 18   *Airfoil Choices*

Designers of modern helicopters have two basic choices for the type of airfoil for their blades: laminar-flow or supercritical. The choice depends on the objectives of the helicopter. If good performance in hover and cruise flight is the uppermost consideration, then the laminar-flow airfoil should be considered. If maximum speed is the design driver, then the supercritical airfoil may be the best choice.

Laminar-Flow Airfoils

During World War II, NACA used their wind tunnels to develop a family of airfoils that had significantly lower drag than previous shapes. The design involved placing the maximum thickness well back from the leading edge. This keeps the air accelerating for a long distance and encourages the boundary layer to remain laminar before it makes its natural transition to turbulent with its higher skin friction thus the designation. These are known as 6-series airfoils in the NACA system of designation. One of them, the 64-412, with its distinctive drag characteristics is shown in Figure 18-1.

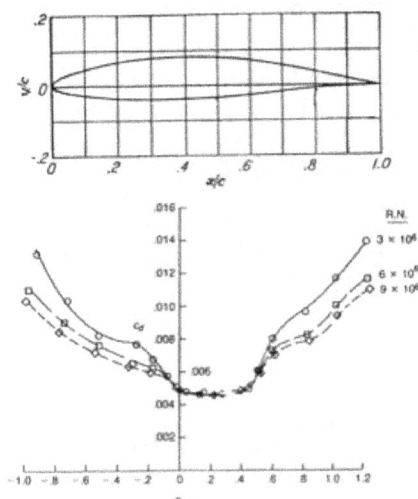

FIGURE 1. A Laminar Airfoil

The lift coefficient range in which this desirable low drag condition exists is somewhat limited, but it can be made to coincide with the wing lift coefficient for airplanes at an airspeed range including cruise speed. Airplane designers immediately picked up on these airfoils with their drag bucket. The North American F-51 and the Lockheed F-80 were among the first airplanes to use these new airfoil sections.

But careful flight testing was disappointing. The benefits that had been measured on smooth models in the wind tunnel did not seem to apply to actual wings. It was discovered that bug spots, rivet heads, and skin mismatches caused early transition from a laminar to a turbulent boundary layer, and thus the drag was no lower than on the older airfoils.

At one point, NACA took a fully instrumented F-51, and applied body putty to make a smooth wing surface that approximated the wind tunnel models. The wing was then wrapped with a plastic sheet and the pilot climbed to 10,000 feet before pulling a lanyard that discarded it. For ten minutes, the power required was low enough to indicate an extensive area of laminar flow, but soon after that some high-flying bugs had nullified it. (I heard that story somewhere, don't ask me for a reference.)

The laminar-flow airfoil, however, has found a home in the sailplane community where great care is taken to have smooth wing surfaces by both design and maintenance.

So, how does this apply to helicopter blades? Tests reported by Tanner and Yaggy in the 1966 AHS Forum provided information on the boundary layer on the blades of a UH-1 with a NACA 0012 airfoil (no claim to laminar flow) in hover. The blades were sprayed with a gray organic chemical that evaporated slowly in a laminar boundary layer, but quickly in turbulence. The transition point could be readily identified after a short flight by seeing where the gray color had been left in the laminar region.

A conclusion of the paper is: "The hovering rotor has the ability to maintain laminar flow even with considerable erosion of the leading edge." This indicates that while projections such as bug spots force the transition, scratches and depressions do not. And bug spots get sandblasted off helicopter blades on every take off. Thus the potential of getting appreciable laminar flow on rotor blades is enough to make the designer consider these airfoils. (A secondary result of these tests was the surprising observation that the boundary layer was not affected by centrifugal forces. Indeed most of the surface flow had an inboard drift.)

The airfoils shown in Figure 18-2 as used on the Chinook and the Apache are variations of laminar-flow airfoils. They were selected in the early 1970s.

FIGURE 2. Airfoils for the Chinook and the Apache

The Other Family

Some airfoils tested in wind tunnels have produced surprisingly low drag at certain combinations of moderately high lift coefficients and Mach numbers. This seems to have nothing to do with the type of boundary layer. One comparison is shown in Figure 18-3. The NACA 23015 has drag characteristics that one would expect, but the 66,2-215 does not. The 1945 report could only characterize the latter as being "peculiar."

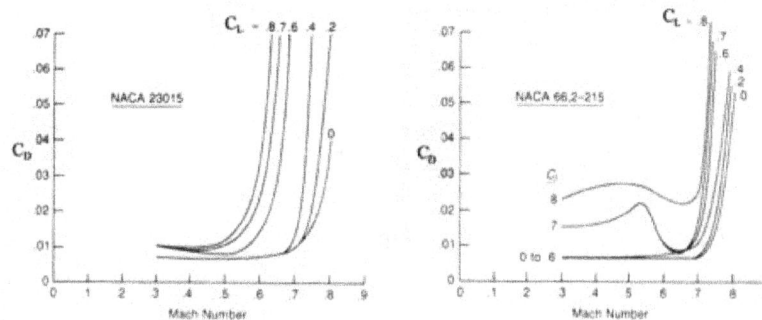

FIGURE 3. Two 15% Airfoils

At the Mach numbers and lift coefficients where the differences exist, the local velocity on the forward portion of the airfoil speeds up until it is supersonic. Air doesn't mind speeding up, but it sure doesn't like to slow down. If it reaches a local Mach number of about 1.4, it will slow down almost instantaneously through a shock wave.

It took some extensive research to understand how the strength of the shock wave at a given test condition is affected by nose shape. For the 66.2-215 airfoil there is a strong shock wave at Mach 0.55 but a weak one at 0.65. Aerodynamicists saw that it was high negative pressure close to the leading edge that indicated the formation of a weak shock wave. This pressure distribution is dubbed "peaky."

The jet transport people now know how to design airfoils so that at the cruise condition the shock is weak, and since the local Mach number is above one, they call these airfoils, "supercritical."

If your airline seat is where you have a god view of the wing and it is pointed either toward or away from the sun, you should be able to see the shadow of the shock wave at about the quarter chord.

Application to Rotors

Helicopter airfoil designers have followed the lead of their fixed-wing brethren to develop super-critical airfoils for rotors. Figure 18-4 shows the desirable characteristics as pointed out by Leo Dadone. These airfoils are characterized by short noses and slab sides. The primary objective is to be able to go to a high angle of attack at a Mach number of 0.4 which is characteristic of the retreating tip.

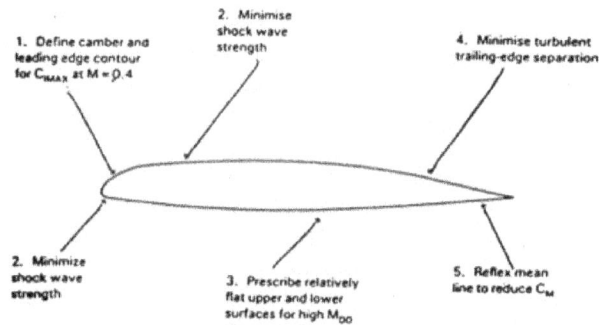

FIGURE 4. A Modern Airfoil Shape

The drag benefits we might get from a supercritical airfoil on a rotor blade are probably not of much importance since they only exist for certain combinations of Mach number and lift coefficient. The effects do come into play, however in the ability to go to a higher angle of attack on the retreating tip before generating a shock wave strong enough to produce retreating blade stall.

Thus if the supercritical affects exist, the helicopter can fly faster than with other airfoils, but because its maximum thickness is reached close to the leading edge, there is little chance for a large area of laminar boundary layer and thus no potential for delivering the potential lower power benefits of the laminar-flow airfoil at hover or cruise. The airfoil selected for the Comanche was supercritical.

The other effect is due to delaying the nose-down pitching moment as the pressure behind the shock wave changes because of shock-wave separation. Since our blades are long and limber and not every blade as the same torsional stiffness nor weight distribution, it is certain that one blade will twist nose-down before its mates. This will change its aerodynamic characteristics and make it fly out-of-track while depositing a different vortex trail for the other blades to contend with. At this point, each blade will find its own peculiar track and the helicopter will shake. This is the primary symptom of retreating blade stall.

CHAPTER 19   *A Look at Low Drag Hubs*

At the October, 2008 Specialists Meeting in Dallas on "Technologies for Next Generation of Vertical Lift Aircraft', a paper presented by the Army concerning advanced rotor technology made a plea for low-drag rotor hubs.

We have heard this many times, but by looking at most recent hub designs, it appears to have been made in vain. Unfortunately, the aerodynamicist generally has less influence than those of other disciplines in the hub design process

The Bad News

When visiting Sikorsky a few years back I was told that the S-92 hub was designed for long life and that its drag was the same as an entire S-76.

FIGURE 1. **The Sikorsky S-92 Hub is Big but Long-Lasting**

The drag of the Comanche's hub was high, not so much because of the shape, but for its very big size dictated by its composite design. (I kept telling people it was a high-drag hub and finally they canceled the project.)

FIGURE 2. The Comanche with a High-Drag Hub

The Good News

Over the years, however, we have seen a few low-drag hubs. Bell led the way by designing the Model 540 "door-hinge "rotor first used on the UH-1C and later on their AH-1G and S. The drag is low because of the flat configuration. Compared to other rotors, it also benefits aerodynamically by the lack of in-plane dampers since two-bladed rotors do not have lead-lag hinges. The oscillating in-plane moments are handled by sufficient structure in the hub as well as a relatively flexible mast.

FIGURE 3. The Bell AH-1G with the "Door-Hinge "Hub

Lockheed adapted the door-hinge design for the four-bladed Cheyenne. The photo shows the rotor with the gyro that gave good handling qualities. If the aircraft had gone into production, that gyro would have been replaced by one under the transmission running at high speed. Late tests in the program showed that this was the preferred configuration. Here again, there were no lead-lag dampers to produce drag. The in-plane moments were handled with stiff titanium blade arms.

FIGURE 4. The Lockheed Cheyenne with a Door-Hinge Hub

Another low-drag hub was proposed by Sikorsky at the 1984 AHS Forum as part of its composite, bearingless "Dynaflex Rotor. "A feature of the cambered, elliptical shape is that aerodynamically it has a negative angle of attack for zero lift so that the hub download is avoided at normal nose-down cruise attitudes. This good idea was never used.

FIGURE 5. Sikorsky Proposed this Low-Drag Hub in 1984

So it is still possible that future helicopters may benefit from low-drag rotor hubs

CHAPTER 20   *Pylon Drag and Tail Shake Remedies*

Question

What are those strange shapes just below the rotor on a number of current helicopters? They are there to reduce drag or to cure "tail shake "

In the Beginning

The first ones I know about were on the McDonnell XV-1 Convertiplane as those bulges on the pylon just under the rotor that can be seen in Figure 20-1.

FIGURE 1. Drag Reducers on Pylon of McDonnell XV-1

They were developed in 1951 in the wind tunnel of the University of Washington at Seattle where I was working part time as a graduate student when McDonnell brought in a powered model of the design. I watched as the model technician formed the "ramps "out of red wax under the direction of Dr. Kurt Hohenemser. The object was to prevent flow separation at the back end of the pylon and thus minimize drag. (Dr. Hohenemser later was a great influence on the American helicopter industry as a professor at Washington University in Saint Louis, Missouri specializing in helicopter dynamics.)

Also at Sikorsky

The next application was also developed during wind tunnel testing; in this case, for the Sikorsky S-61. The configuration shown in Figure 20-2 was again to reduce drag and was on the mock-up built at the beginning of the project so it was not an after-thought as many others were.

FIGURE 2. The Pylon Fairing on the Sikorsky S-61

The explanation of its purpose is that it generates tip vortices as a low-aspect ratio "wing "that energizes the boundary layer on the sides of the pylon and keeps it from separating and thus producing drag. The idea is illustrated in Figure 20-3.

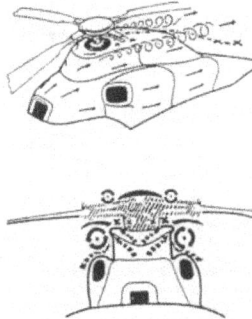

FIGURE 3. Vortex Patterns Generated by Pylon Fairings

Besides the S-61, quite a number of helicopters have a similar installation including the U.S. Coast Guard's HH65A, the Mil Mi-26, the Sikorsky Black Hawk, and even on the front pylon of the Boeing Chinook.

FIGURE 4. **Pylon Fairings**

Something to Think About

There are opposing considerations for such a device. Preventing flow separation reduces drag, but the fact that vortices are being generating means that energy is being put into the flow and this results in "induced "drag. Thus the aerodynamicist should be aware of the net effect of the installation.

Another Device

Figure 20-3 also shows a fairing on top of the rotor hub. The idea for this came from a Sikorsky project to use fluid to de-ice the rotor blades as was done on propellers at the time. To examine the aerodynamic effect of a de-icing tank mounted on the rotor hub, a wooden mock-up was flown. There seemed to be no obvious problems with the configuration, but the pilot made a comment. He said, "That was the smoothest ride that I ever had in this bird. "After the next flight without the tank mock-up, he reported, "It shakes just as much as it always did. "Now you see a number of helicopters sporting a "beanie "to reduce vibration.

The Source of Chaos

The flow field at the rear end of the helicopter has been affected by the rotor and also by the air's passage past the fuselage, landing gear, and engine nacelles. Thus the environment at the rear end can only be described as chaotic. It is no wonder, therefore, that the empennage and/or tail rotor can react to this flow by producing erratic forces that shake the entire aircraft.

Both devices, the one below the rotor and the one above it, are generating vortices which tend to organize the flow into a somewhat lesser chaos and thus reduce tail shake. When the problem shows up during early flight tests, they are often installed. Figure 20-5 shows this fix on the Bell OH-58D when flight test revealed a tail shake problem that the Jet Ranger--from which it came--did not have.

The Bell Jet Ranger   Its Son, the OH-58D

FIGURE 5. A Configuration Change Made after Flight Test

In some cases the problem can be traced to some parts of the helicopter having structural natural frequencies that match the input frequencies of blade passage of the main or tail rotors. This gives the dynamics engineer a chance to change natural frequencies and avoid resonances by stiffening structure to raise a natural frequency or softening structure or adding weight at strategic locations to lower it

CHAPTER 21   *The Offset Flapping Hinge*

When Jaun de la Cierva invented flapping hinges to make his autogyros flyable, he decided to place them as close to the center of rotation as possible. Subsequent designers of autogyros and early helicopters followed his lead. If you see a Sikorsky R-4, R-5 (S-51) or R-6 in a museum, you will notice a rather complicated mechanical arrangement of the hub to accomplish this. When, in the late 1940's, the Sikorsky S-52 (a four-place utility helicopter) was coming along, the hub designer said, "I know how to make a simpler configuration. "That's the start of the off-set flapping hinge.

Sikorsky test pilots who had been flying those earlier helicopters were pleasantly surprised by the increased control power of this new design. Engineers had to think a bit before they came up with the explanation. It was that rotor flapping did not only produce a moment about the Center of Gravity due to the tilt of the thrust vector but it also produced an effect due to the fact that the centrifugal forces acting at the hinges did not line up with each other thus adding a "couple "at the hub as shown in the sketch. This adds to the effect of the tilt of the thrust vector to produce moments about the Center of Gravity.

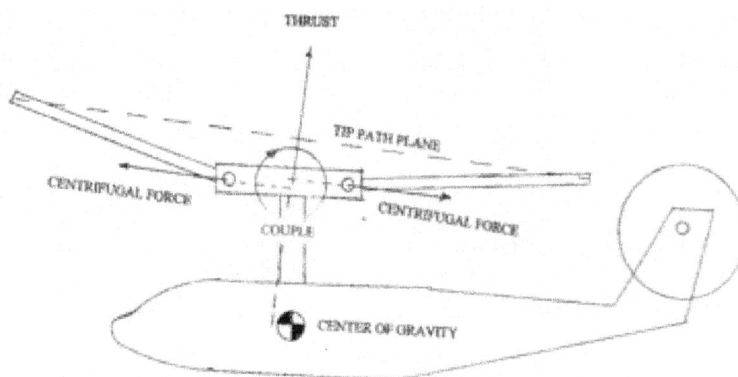

FIGURE 1. **Generation of Hub Couple**

Designers of rotors with more than two blades quickly adopted this design. The measurement of how much effect is achieved with offset mechanical flapping hinge is related to the percentage of offset with respect to the blade length. Values of two to five percent are typical.

One thing that happens with offset flapping hinges is that while the centrifugal force on a local increment of mass is a function of its distance from the center of rotation, the flapping moment of inertia of this element is a function only of its distance from the hinge. From a dynamics stand-point, this changes the flapping natural frequency from being exactly equal to the rotational frequency (a system in resonance) where the response to a control input is exactly 90 degrees later to something slightly less.

This produces "cross-coupling. "If the pilot pulls the stick straight back to make maximum cyclic pitch on the right hand side of the disc, he will get the nose-up response he was expecting, but also a bit of left roll since the flapping is maximized just to the right of the nose. Pilots rapidly learn to compensate for this.

In the late 1950s, a new rotor design was introduced - the hingeless rotor. Although earlier designers, such a Stanley Hiller, had experimented with it, the Lockheed "Rigid Rotors "were the ones that led the industry in this direction. In these helicopters, cyclic pitch is used to balance the aerodynamic forces for steady flight and to also introduce enough additional pitch to bend the blades to make the rotor produce moments to balance moments existing on the airframe such as those due to center of gravity offsets, fuselage aerodynamic moments, horizontal stabilizer loads, and moments due to tail rotor thrust and torque. For maneuvers, additional cyclic pitch is imposed on the trim values to produce precession rates on the rotor as on a gyroscope to do pitch and roll maneuvers.

Since there are no actual hinges, engineers have to come up with an "effective hinge offset "to do analysis with these rotors. The dynamic engineers calculate the natural frequency compared to the rotational frequency. Knowing that, they use an equation that produces the effective hinge offset. The flying qualities engineer can then use this as if it were the location of a real mechanical flapping hinge. These values are much larger than those associated with mechanical hinges - eight to fifteen per cent of radius. The cross-coupling can be such that an aft stick motion produces maximum flapping only 75 degrees later than maximum cyclic pitch.

Pros and Cons

A helicopter with a hingeless rotor has a very quick response to control inputs. The technical term is that it has a low "time constant. "Pilots say the control response is "crisp. "The helicopter with the lowest time constant is the BO-105 which is preferred for performing at air shows.

Mechanical flapping hinges using bearings are subject to high centrifugal loads and small oscillating motions. These effects can produce galling unless constantly lubricated. You can see the oil containers on the hubs of Sikorsky helicopters built before the Black Hawk and S-76. On these recent helicopters, the bearings have been replaced by elastometric material that provides the ability to flap, lag, and pitch without needing lubrication. Of course, the hingeless design also is free of this complication.

Although the hingeless rotors have some good points, they also have some bad ones. One is the high cross-coupling mentioned above. On the BO-105, with fiberglass blades cantilevered from a very rigid hub, when the pilot pulls the stick straight back and gets maximum pitch only 75 degrees later, the rolling acceleration is more than the pitch acceleration! This is due to the aircraft's lower moment of inertia in roll than in pitch. New BO-105 pilots have to learn to compensate for this surprising effect.

Another consideration is that if the cyclic stick is moved while the aircraft is on the ground, the resulting bending of the blades will produce high hub moments; in some cases enough to roll the helicopter over. The Lockheed Cheyenne had sensors in the landing gear that limited the cyclic pitch when the oleos were compressed. The BO-105 has a strain gauge system in the mast that activates a warning in the cockpit before the bending moment reaches a dangerous level.

Another phenomenon applies to all helicopters, but more so to those with hingeless rotors. When making a fast stop or a landing flare from a fast descent, the tip vortices are initially left behind and above the rotor plane. At the end of the maneuver, however, when the helicopter is in hover, the tip vortices are going down below the rotor. Thus at some point during the transition, the tip vortices are in the plane of the rotor producing a chaotic aerodynamic environment.

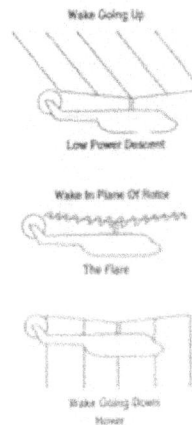

FIGURE 2. Conditions in a Steep Landing Maneuver

No matter how the blades are attached to the hub, they will flap and bend about the same amount during this time of chaos. The flapping and bending will produce moments in the hub which result in vibration in the airframe, but the stiffer the rotor, the higher these moments will be. A pilot who flew the BO-105 servicing oil platforms told me that he has vivid memories of this heavy vibration.

Bernhard Enenkl of Eurocopter Deutschland has informed me that the effective hinge offset of the BO-105 is 13.6% (quoted as 15% in some publications). Apparently the experience with the BO-105 prompted the design team of the BO-108 (now the Eurocopter 135) to use a different hub design with an effective hinge offset of only 8.4%. Even those helicopters using the BO-105 hub, the BK-117 and the Eurocopter 145 (the U.S. Army's Lakota), have somewhat softer rotors with effective offsets of 12.0% and 11.8% respectively.

21

The Offset Flapping Hinge

CHAPTER 22 — *A Noise and Weight Trade-off Study*

It seems that for every good thing we can do to a helicopter, there is a price involved. This is true for making the helicopter quieter by reducing its rotor tip speed. The empty weight will go up. Most of this is in the transmission which is delivering the same horsepower to the rotor at a higher torque. Since the main rotor torque is higher, the weight of the tail rotor will also increase.

I have used the example helicopter of my textbook to determine the weight penalty. In the case of this design, the tip speed of 650 feet per second is already low enough to classify this helicopter as quiet (I was working on the Lockheed Cheyenne at the time and that was its tip speed). So I have gone the other way for this study by going from 650 to 750 ft per second thus obtaining a saving of empty weight while making more noise.

Weight equations for various airframe components are listed in my textbook. They were taken from Army reports of the 1970's, but I have recently been told that they are still good.

The study was made with the following ground rules:

1. The gross weight is 20,000 pounds and any saving of empty weight goes into increasing the payload.
2. The blade loading coefficient, $C_T/2$, remains the same. (By increasing the tip speed while holding a constant blade loading coefficient, the required chord is reduced from 2 feet to 1.73.)
3. The rotor polar moment of inertia remains the same

The results of using the equations are shown in the table.

Component	Original Weight	New Weight (Tip Speed = 650 fps)	Weight Saving (Tip Speed = 750 fps)
Main rotor blades	830	790	40
Main rotor hub	636	676	- 40
Tail rotor	180	159	21
Drive system	1673	1526	147
		Total weight saving	168

The saving of 168 pounds would also be the weight penalty had the original tip speed been 750 feet per second and was being reduced to 650. Based on experiments with actual helicopters, the noise level would be significantly reduced with the lower tip speed, but you would have to leave one passenger behind.

This applies to reducing external noise, but reducing internal noise is also a desire. One criteria is to provide people four feet apart with the ability to carry on a normal conversation. This requires the installation of insulation and possibly noise-canceling systems at the sacrifice of several other passengers. This is the kind of trade-off that helicopter designers constantly face

CHAPTER 23   *Fly-By-Wire*

In the last twenty years, helicopter control systems have evolved from simple mechanical collections of push-pull rods and bell-cranks into much more sophisticated systems. The ultimate as used on the V-22 Osprey and the lamented Comanche is the fly-by-wire system.

What We have Today

Before going into those helicopters, let's look at Figure 23-1 as a simplified sketch of the longitudinal control system used on the Boeing Apache. It is a collection of push-pull rods and bell-cranks but with some special features. For one, it has a dual hydraulic system to help the pilot move the swashplate for changing cyclic pitch. Many small helicopters get away without this feature, but with helicopters as big as the Apache and the Black Hawk, the cockpit control forces would be too much for the pilot to easily handle. One hydraulic system might be adequate if it could be guaranteed that it would never fail, but without that guarantee, a second is desirable.

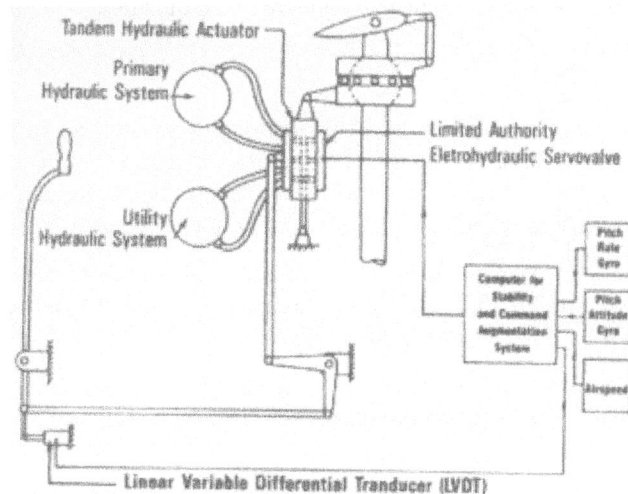

FIGURE 1. **Apache's Fly-by-Iron Dual Hydraulics and SCAS**

The pilot is moving a simple valve on the hydraulic actuator that tells the system which side of the piston needs pressure to respond to his command. (Note that in this application, the piston is fixed to structure and so it is the cylinder that moves the swashplate.) Since the force to move the valve is small, a spring is attached to the stick to give the pilot a feel of what he is doing.

The other addition to the system is a computer, getting signals from a pitch attitude gyro, a pitch rate gyro, and an airspeed sensor. This is the basis of the Stabilization Augmentation System (SAS). Working with the inputs from the sensors, the SAS improves flying qualities by adding damping, improving turn coordination, tailoring inputs, and generating attitude hold and hover position hold when the pilot selects them.

If the helicopter gets upset by a gust, the SAS sends an electrical signal to an electro-hydraulic valve on the side of the actuator to allow pressurized hydraulic fluid to move the cylinder. Thus, the SAS does the right thing to bring the aircraft back to its original flight condition without the pilot having to. To prevent a computer failure from producing an actuator "hard-over, "the authority of the computer is limited to about ten per cent of full actuator travel. Even the effect of a 10% hard-over is minimized by monitoring the computer and nulling its command if it is detected doing something dumb.

Since the computer can not distinguish between a pitch response produced by a gust and a response that the pilot is calling for in a maneuver, there must be a means for taking care of this. It is done with a wire from the cockpit carrying an electric signal that tells the computer to ignore the gyro inputs when the pilot moves the stick. Thus we already have a partial fly-by-wire system. The addition of this wire can be said to convert SAS to SCAS (Stability and Control Augmentation System)) and the signal can be used to modify the pilot's command by using the computer to speed it up or slow it down depending on different flight conditions.

Note that this system has the capability of accepting signals from any type of sensor, processing them in any manner, and using them to improve flying qualities.

Another feature

The Apache also uses this wire for a Back-Up-Control System (BUCS). In case the mechanical control system becomes disconnected due to combat damage, a motion of the stick without a corresponding motion of the swashplate will immediately convert the computer outputs to control 100% of the actuator travel and the pilot can continue flying as before.

A jam in the control system can also be fixed by using enough force on the control to break a shear pin to disconnect it. In this case, the conversion is not immediate but ramps up in three seconds to prevent a sudden extreme control input.

Continued flying for any length of time while depending on a one-wire-control system is not recommended, so the pilot should land, or with luck get back to his base.

The next step

An obvious extension of BUCS is to use the wire from the stick to a 100% authority computer to eliminate all push-pull rods and bell-cranks thereby hopefully save weight. For safety, of course, all elements have to be triply redundant as shown in Figure 23-2. As part of the system, a monitor compares the signals from all three computers and if one is not the same as the other two, it will disconnect it.

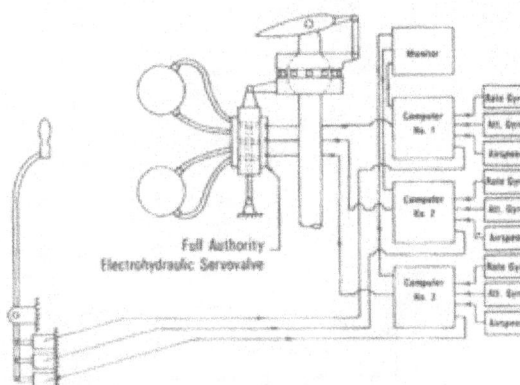

FIGURE 2. Triply Redundant Fly-by-Wire System

Considering the monitor, the two extra computers, the additional sensors and all the wire bundles that connect them, it is not clear that fly-by-wire would provide a significant weight saving.

One possibility of the new system is that the central cyclic control stick can be replaced by a "side-arm controller. "It produces electric signals as the pilot applies forces and/or displacements. An advantage is that it gives the pilot a better view of his instruments and control panels.

FIGURE 3.

FIGURE 4. **Side-Arm Controller**

A paper from the 2007 AHS Forum by Wittmer, Knaust and Stiles describes the fly-by-wire system being developed for the Sikorsky S-92F. Because of the side-by-side pilot location, Sikorsky has decided not to use side-arm controllers but to use centrally-located, small-displacement versions of them. The collective levers and pedals are also small-deflection controls that produce electrical signals.

Fly-by-wire proponents list features that their systems can have, but many of them can also be accomplished by the simpler Apache system. Another 2007 Forum paper by Jeff Harding proposes a modified computer for the AH-64 D that would satisfy most of those fly-by-wire goals.

My Opinion

Based on what I know about the Osprey and Comanche fly-by-wire programs, if I were the chief engineer of a helicopter company faced with designing a new machine, I would tell my engineers, "No fly-by-wire "

If you--as my competitor--did use fly-by-wire, I would have my aircraft in service three years before yours, at half the development cost and at two thirds the fly-away-cost.

When you finally did get yours into service, it would do a few things mine wouldn't, but the question would remain; "Was it really worth all that extra effort?'

Besides that, you would still be faced with years of potential hardware and software problems. Some might occur quickly as happened with the fly-by-wire Osprey, Ship 5, which crashed on its first takeoff. Or it may take a long time as in the case of the billion-dollar fly-by-wire B-2 bomber that crashed in 2008 because moisture got into its airspeed system.

Imagine an Apache crew chief who wants to check the control system. Upon opening up the ship, he will see bellcranks, rods and other mechanical parts joined together and will have a pretty good feeling that the system is working as it should. On the other hand, the crew chief of the fly-by-wire Osprey will see wires and black boxes and it will not be obvious that the system is working as it should. He has to use an elaborate testing system whose development was probably a contribution to the schedule slippage and cost over-run of this project. And even then, he may not be able to guarantee that everything is O.K.

Certainly the B-2 bomber has a similar testing system, but it was unable to find the moisture in the airspeed system that was the cause of the destruction of one of these very expensive aircraft in Guam in 2008. With hindsight, we can say that the engineers should have foreseen this possibility.

Reasons for My Opinion

The above is my column in the Spring 2009 issue of Vertiflite giving my concerns about fly-by-wire control systems for helicopters. It produced two responses.

Nick Lappos, the former Comanche test pilot now at Bell, told me that while what I had written was true, he thinks that the advantages of the fly-by-wire system outweighs the substantial additional time and expense needed to put it into operation.

Mark Tishler and Chris Blanken of the US Army Research Center at Moffett Field. California were more specific in their letter printed in the Fall 2009 edition of Vertiflite while agreeing with Nick and disagreeing with me

One of their points is that the difference between the control systems as used on conventional helicopters like the Apache and Black Hawk and new aircraft like the Comanche and Osprey is one of authority. On an old system the single computer with its gyros and other inputs is limited to using only plus-or-minus 10 or 15% of full actuator travel. This assures that a computer malfunction will not make a big control input.

Tischler and Blanken point out that even if helicopters with "partial-authority "systems can have good flying qualities, meeting the requirements of the Airworthiness Design Standard (ADS-33) which was the flying qualities specification written specifically for the Comanche, the fly-by-wire system with full authority provides some opportunities for flight characteristics not possible with the simpler system.

Fly-by-wire Accidents

A Black Hawk accident at the Sikorsky plant in Stratford was due to a mechanic forgetting to hook up the lines from the airspeed system to the stabilator computer. During an aggressive takeoff, the stabilator stayed at its high-incidence hover setting and the unexpected large nose-down pitching moment forced the helicopter into the river.

The first-flight accident on the V-22, ship 5, was due to the assemblers making two identical wiring errors in two of the three computers. When the monitor saw one computer being out-of-step with the other two, it was shut down, but it was the only one doing the right thing and the pilot lost control without it.

The other big V-22 accident was due to some wires of the fly-by-wire system chafing against hydraulic tubes and causing a leak.

Since these mistakes are well-documented, they will probably never happen again, but there must be a few of their cousins still lurking in these full-authority systems.

Tischler and Blanken agree with me that the weight saving of installing fly-by-wire on "legacy "helicopters (another way of saying "old') is "probably a wash, "but they predict that there is a possibility, by using load-limiting features with fly-by-wire in new designs for "considerable structural weight saving since the aircraft no longer has to be stressed for full mechanical control input at any flight condition "

I believe that this is questionable. Once the airframe is designed for hard landings and made "crash-worthy, "it should not have to be further strengthened for flight loads. Helicopters are different than airplanes. At high speeds, airplane wings can take advantage of the high dynamic pressure to produce high lift and therefore high stresses. Rotors, on the other hand, cannot take advantage of the high dynamic pressure on the advancing blade since its lift cannot overpower the lift of the retreating blade where the dynamic pressure is low. (Not considering Sikorsky's ABC aircraft.) Most of the lift in level forward flight is generated by the blades over the nose and tail, and they think they are still in hover with angles-of-attacks about half-way to stall Thus the ability of a helicopter to produce high load factors during maneuvers is limited to two or three G's.

The structure of U.S. military helicopters must be designed for a 3.5 load factor (think of battery supports) although such a high load factor does not have to be demonstrated in flight test.

There have so far been no requirements for really high load factors. The Utility Tactical Transport Aircraft System (UTTAS) competition which Sikorsky won with the Black Hawk had a required maneuver during which the helicopter at 150 knots had to pull up to 1.75 G "s. The prototype Black Hawk could not do that until the blades were given new airfoils.

The above discussion applies to the faint possibility of reducing airframe structural weight due to steady loads, but does not address oscillatory fatigue-causing loads such as hub moments. Proponents of fly-by-wire speak of its ability to provide "carefree handling "by automatically limiting control displacements or at least by warning the pilot of high loads. (Of course since it is only fatigue life that is of concern, in an emergency, the pilot must be able to over-ride any limit that the control system is trying to maintain.)

I can see that this might be useful in an existing helicopter, but in a new one, it is asking the controls engineer to somehow compensate for the poor estimates made by the loads and stress engineers. In many cases, a few extra pounds of material in the original design would have given inherent carefree handling.

An example of this was the transmission of the original Bell Cobras. During an entry into a left turn, an aerodynamic phenomenon requires high engine power that the pilot sees on his torque meter. During the air-to-air combat trials at Patuxent River in the 1980's, Cobra pilots were reluctant to make left turns because of this. Thus they were respecting a "limit. "This type of limit should not be a candidate for a fly-by-wire application. Bell did the right thing by upgrading the transmission. (The Russians avoid this problem by not installing torque meters in their helicopters.)

Now you see some of my concerns besides long development time and cost of fly-by-wire control systems for helicopters. I expect to hold onto this opinion for the rest of my life; and I'm only 83!

Fly-By-Wire

CHAPTER 24   *Another Look at Igor's VS-300*

My first column for Vertiflite appeared in the Fall-Winter 2000 issue and concerned the fact that four pilots in the 1940's could fly the Sikorsky VS-300 after just a few minutes of instruction when it had two horizontal tail rotors as shown in Figure 24-1.

FIGURE 1.

FIGURE 2. The Sikorsky VS-300 with Horizontal Tail Rotors

At the time, I attributed this ease of flying to the large damping in pitch and roll generated by those two horizontal tail rotors. I now believe that I missed what may have been even more important: the increase in thrust of these rotors as they were subjected to translational velocity.

How it is Now

It usually takes several hours of dual instruction before a modern helicopter student has mastered hover. The reason is that the helicopter he is trying to fly is dynamically unstable in hover. This is due to the flapping of the main rotor as the helicopter moves from its hover position.

If the aircraft is somehow upset--for instance nosed down by a gust--and the pilot refrains from controlling it, it will start moving forward from its hover position. The resulting aft flapping will produce a nose-up pitch rate and a nose-up attitude that will stop the forward motion, but the attitude when it stops will be more nose up than the original nose-down displacement which started the maneuver. Thus after coming to a stop, the helicopter will go backward with more energy than it had initially and will oscillate with ever increasing amplitude. This is the sign of an unstable system.

I once described this instability in a company class I was teaching. One of the students was a very experienced test pilot and he disagreed with me. He invited me to see how wrong I was by flying with him. He trimmed up in hover and made a very small twitch with the lateral control and then froze. The helicopter started moving to the right but in several seconds it stopped and started to the left where it again stopped but with more bank than it had when it started and then it repeated the cycle at higher and more exciting amplitudes until my friend, the very experienced test pilot, suddenly took control. He turned to me and said, "Ray, you must be right "

Most student pilots are not told of this characteristic, but by trial and error they start learning to control the instability right from the first. The period of oscillation depends on the size of the helicopter varying from about 13 seconds for the Robinson R-22 to 24 seconds for the Sikorsky CH-53E. The time to double in amplitude depends on the height of the rotor above the center of gravity and the offset of the flapping hinges. The larger these values are, the more unstable will be the helicopter in hover.

At cruise speeds, however, this effect is stabilizing. If the helicopter increases speed from its trim condition such as inadvertently starting a dive, the aft flapping will pitch the helicopter up and slow it down to its original speed. We call this "speed stability "

(An interesting sidelight of this explanation is that if the center of gravity is above the rotor, the system is stable as has been demonstrated by the DeLackner "stand-on "helicopter of Figure 24-2. It could be controlled by simply leaning in the direction the pilot wanted to go.)

FIGURE 3. The DeLackner Helicopter

The VS-300 was Different

Why then, was the VS-300 different? Of course we know that the main rotor was flapping to produce instability, but the other two rotors were increasing thrust with speed and with their large moment arms about the center of gravity, they were counteracting the instability of the main rotor. Thus the four pilots who could fly the VS-300 with very short instructions did not have to be stabilizers themselves. (It should be pointed out that the stabilizing effect did not exist for sideward flight as it did for forward, but the high roll damping helped the pilots)

Those Pilots

You can read about the first two outside pilots in Pioneering the Helicopter by Les Morris and Helicopter Pioneering with Igor Sikorsky by William Hunt. They were Captain Frank Gregory and his second-in-command, Lieutenant Victor Haugen of the Army Air Forces at Wright Field, both being autogyro pilots. In July of 1940, they had been inspecting the yet-to-fly Army-sponsored Platt-LePage helicopter project near Philadelphia and decided to look at what Sikorsky was doing on his own at Bridgeport on their way home. Igor saw a good public relations opportunity by inviting them to fly the VS-300. Since it had only one seat, Igor instructed by standing on the tarmac while telling them what those levers were for as illustrated by Figure 24-3.

FIGURE 4. Captain Gregory in Ground School

With only five minutes of instruction, Captain Gregory took off and like any new pilot had an exciting first ride while learning the machine's characteristics. Les Morris says. "It bobbed around like a toy balloon in a gale. "But Gregory went up again and in ten minutes could hover and fly slowly around the field. Figure 24-4 shows Gregory in full control of the VS-300. Lt. Haugen then had his chance, and according to Hunt, learned even faster than his boss. (During the war, both pilots were given fast promotions with Gregory winding up as a General. He was responsible for Sikorsky getting a contract to develop and produce the R-4.)

FIGURE 5. **Captain Gregory after Ten Minutes of Trying**

The third new pilot a few months later was Charles Lindbergh, who repeated the experiences of Gregory and Haugen. The next was Les Morris. Igor and his executive engineer, Serge Gluhareff, had done all the initial flying, but as the project became more involved, they felt that they needed a back-up so they hired Les as the first full-time test pilot. On his first flight, he also got a lot of wobble as Gregory had, but he writes, "After numerous trials, each one lasting perhaps a second or two longer than the one before, I was able to make appreciable flights "

A Suggestion

If it is desirable to have a helicopter that is easy to fly, perhaps the Sikorsky VS-300 configuration with two horizontal tail rotors should be considered.

CHAPTER 25   *An Easy-to-Fly Helicopter*

A letter published in the Spring 2006 issue of Vertiflite from Tom Hanson recommended paying more attention to improving flying qualities.

Tom has a special interest in this subject since he was one of a small group of engineers who in the 1950s designed Lockheed's first helicopter, the CL-475 (the 475th preliminary design study at the California Lockheed facility). Figure 25-1 shows this aircraft.

FIGURE 1. The Lockheed CL 475, Hovering Hands-Off

The key to its very good stability was coupling slightly swept-forward blades with a gyro. The gyro concept was not too different from that used by Bell and Hiller in their early days, but the use of hingeless blades instead of a teetering hinge made a big difference. (Note: early talk about this system referred to it as a "Rigid Rotor, "but the blades did have some flexibility. Today we would call it a "Hingeless Rotor "with a large effective hinge offset. It was similar to those on such later helicopters as the BO-105.)

Analysis of the hovering characteristics from the Equations of Motion using the gyro as a separate "Degree of Freedom "responding to feathering moments from the swept-forward blades revealed a highly-damped, short-period oscillation rather than the negatively-damped, long-period oscillation that student pilots today require several hours of trial-and-error to learn to live with. Pilots with no helicopter experience could fly the CL-475 after just a few minutes of orientation. This is because the rotor-gyro system was taking care of those unstable pitch and roll characteristics and the only thing the pilot had to learn to work with were the collective pitch and tail rotor controls.

The control system, shown in Figure 25-2, looked similar to that of a conventional helicopter except that the control linkages contained springs. This was because the gyro, which was attached to the swashplate, had to be "precessed "with a force to produce cyclic pitch. As a gyro, its response was 90o later than the input. When the gyro had precessed to give the cyclic pitch the pilot wanted, he relaxed on the spring force by neutralizing the stick and the gyro stayed at its new position. (This meant that the CL-475 did not have a sense of "speed stability as indicated by the trim longitudinal stick position. Later versions of Lockheed helicopters used a horizontal stabilizer carrying a down-load to remedy this short-coming.)

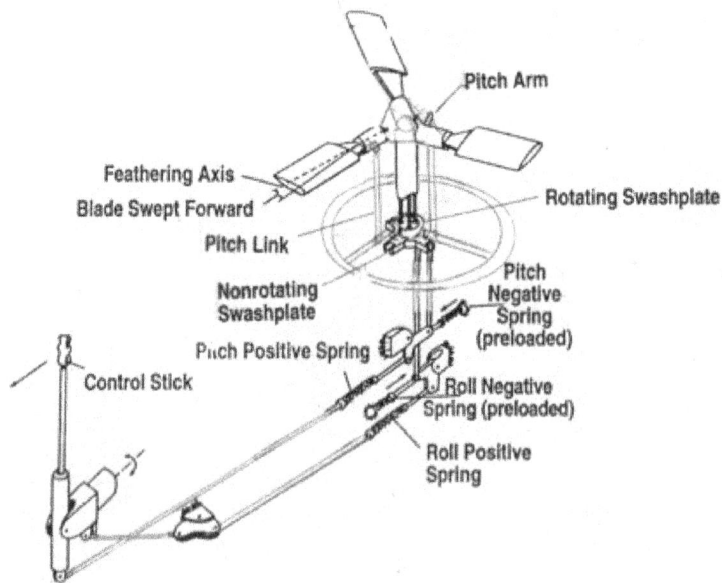

FIGURE 2. The Control System

Because of the sweep forward of the blades and the resulting feathering moment applied to the gyro when they responded, the system wanted to fly with no blade moments. But if some flapping was required to balance such things as an offset Center of Gravity or aerodynamic pitching moments on the fuselage, the pilot could trim by using the spring in the control system to apply just enough force to the gyro to hold the cyclic pitch against the feed-back moment from the bending blades.

The gyro was also used for stability. If, due to a disturbance from outside sources, the helicopter pitched or rolled, the gyro, being a gyro, wanted to stay put with respect to the horizon. This produced a displacement between the gyro and the shaft which resulted in a cyclic pitch input that made the tip path plane come parallel to the gyro. The fuselage followed, thus providing stability. A simple analogy is presented in Figure 25-3.

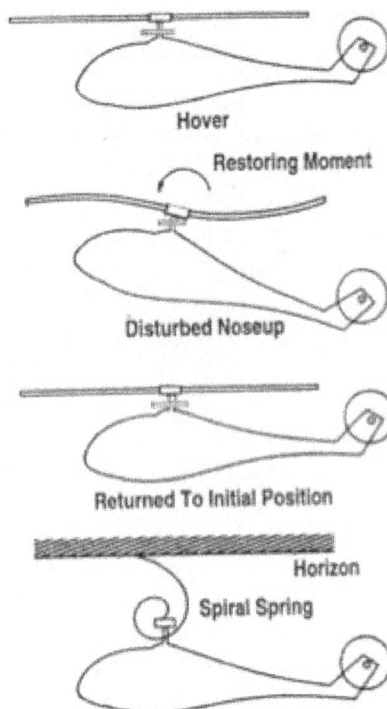

FIGURE 3. A Simple Analogy for Short Time Motions

There is an "yes-but "to be considered here. It might seem that if the pilot were holding the stick, the resulting spring force as the shaft moved while the gyro was standing still would force the gyro to leave its fixed position in space. This was overcome by putting a "negative spring "near each positive spring. The negative spring, with a preload, was much like the toggle on a light switch. As a result of the two springs, the gyro did not see a resisting force and was free to do the right thing when it was stabilizing, but was forced to do the pilot's bidding when he moved the cyclic stick.

Testing the CL-475

The CL-475 was developed in secret and its early flights were made at a remote base consisting of a tent and a trailer on a far corner of a dry lake near Edwards Air Force Base. Lockheed did not have a helicopter pilot at the time, so a crop-duster was contracted to test the aircraft. One day, several miles from the base, an engine problem resulted in a precautionary landing. The pilot had a job to do and so left. The engine problem was soon fixed and instead of towing the aircraft back to the base, a Lockheed mechanic, who was an airplane pilot, but who had never been in a helicopter, got in and flew it back.[1]

Later, as the project matured and was less secret, pilots from the FAA and the military were asked to evaluate it. They were all impressed with its good flying qualities. After his flight, an Army captain remarked, "If you have a bird that any private or general can fly, who needs me?'

This brings up the question about pilot ego. Once he learns to hover a hard-to-fly helicopter, there may not be much motivation to fly one that is easy-to-fly. It may be like telling a bicyclist that he would have fewer spills if he were riding a tricycle.

---

[1] Another discussion of this subject by Tom Hanson can be found in the Fall/Winter 2000 issue of Vertiflite.

CHAPTER 26   *Hovering Over Rough Ground*

We normally consider ground effect as it affects helicopter performance while hovering over a smooth, solid surface, but what if the surface is neither smooth nor solid?

A Simple Test

This was looked at experimentally many years ago by students at California Polytechnic Institute as a side-study when their man-powered helicopter was being developed. As part of the project, they had built a six-foot diameter rotor to be tested very close to a sheet of plywood to represent their aircraft flying in ground effect. The rotor was mounted on the end of a beam which in turn was mounted on a trunnion several feet away that could be raised and lowered to obtain various rotor heights above the surface. The electrical power to generate enough thrust to lift a fixed weight to a level attitude was measured for each test condition.

At my suggestion, they looked at two other conditions by covering the plywood with "AstroTurf "and also by replacing it with a large tray of water.

Much to my surprise, the measured ground effect with the rough surface with the rotor at a height of one-half its diameter was 30% stronger than over the smooth surface. This is in contrast to what pilots tell me. With the pan of water, the experimenters could measure no change in ground effect compared to the solid surface.

A Better Test

You have to admit that a pilot hovering over bushes or long grass may be confused as to where the surface really is. I propose a simple test that bypasses the pilot's perception. It involves using a ground observer who has a good view of the helicopter and something for reference in the background as in the photograph.

FIGURE 1. Hovering over Rough Ground

A golf course would be an ideal flight location. The pilot would set up a hover in ground effect over the smooth fairway and then let the helicopter slowly drift over the rough. As it crossed the boundary, the observer could see whether it went up, indicating increased ground effect or down, indicating the opposite.

Of course, this could develop into a more complicated program. Using a surveyor's transit and the magic of triangulation, both the initial height and any change could be measured. The test could be expanded by flying at both minimum and maximum gross weights and over roughs with various characteristics.

One of the results could be the establishing of the altitude at which ground effect disappears by increasing the hover height until no change in collective or power is needed to hold an altitude higher than the last point.

I would be interested to hear about the results of such tests.

CHAPTER 27 *Human Powered Helicopters*

The American Helicopter Society has a check for $250,000 waiting for you or your team if you are the first to successfully hover a human-powered helicopter. This comes as a result of the Igor I. Sikorsky Human Powered Helicopter Competition.

The Rules

The flight only has to last for a minute, but at some point during that time, the lowest part of the machine must be three meters (just under ten feet) above the take-off surface.

If you decide to go for it, you will join the many who have designed, a few who have built, but only two (as of 2009) who have actually gotten into flight.

The challenge is to follow the track of the Gossamer Condor human-powered airplane that won a similar prize in 1977. We, however, have an additional consideration. An airplane in forward flight can take advantage of a large mass flow of air, but a helicopter in hover has to generate its own mass flow and thus has to work harder.

First Guesses

A good athlete, such as a bicycle racer, can produce about half a horsepower for a minute. Copying the construction of several human-powered airplanes using light-weight materials such as graphite, balsa wood, and Mylar, we can envision our machine with pilot/engine weighing about 250 pounds. Using momentum theory to match the power required to the power available--even considering ground effect--results in the disc loading being in the neighborhood of half an ounce per square foot.

The Da Vinci

The first aircraft of this type to actually fly was the Leonardo Da Vinci III at the California Polytechnic University at San Luis Obispo. It was the end product of the efforts of a series of students under the guidance of Professor Bill Patterson starting in 1981 and ending with flight in 1989. It had a two-bladed, one hundred foot diameter rotor powered by tip propellers.

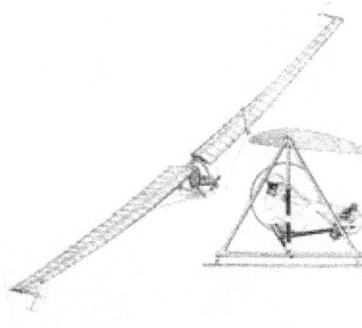

FIGURE 1. Scheme for Da Vinci III

FIGURE 2. The Da Vinci about to Take Off

The pilot/engine provided power by pedaling and winding up strings wrapped around the propeller shafts. The project ended with a flight of eight seconds reaching an altitude of several inches as shown in Figure 27-3.

FIGURE 3. **The Da Vinci in the Air!**

The Next Attempt

The second to fly was the Yuri I at Nihon University in Japan in 1994. It had four two-bladed, 33 foot diameter rotors and was also powered with pedals. (Yuri means "lilly "in Japanese.) The effort was led by Professor Akira Naito.

This aircraft was a little more successful than the CALPOLY attempt, reaching twenty inches in a flight that lasted for twenty seconds. See Figures 27-4 and 27-5.

FIGURE 4. The Four-Rotor Yuri I

FIGURE 5. The Yuri in the Air!

CHAPTER 28   *A Possible Tail Rotor Problem*

There is a maneuver that can produce a traumatic experience for the tail rotor. This was discovered during early flight tests of the Sikorsky HSS-2, the Navy version of the S-61. Stopping a fast right hover turn by stomping on the left pedal resulted in a failure of the tail rotor drive system.

The Navy is interested in having helicopters that are easy to store on crowded hangar decks of aircraft carriers and so Sikorsky designed a hinge at the end of the tail boom so that the empennage could be swung around to reduce the storage footprint.

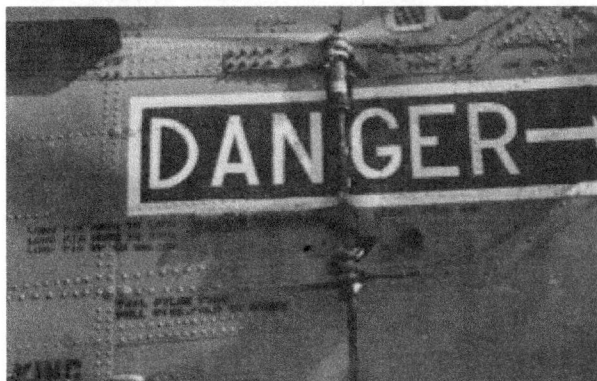

FIGURE 1. The Hinge for Folding the HSS-2 Tail Pylon

At the hinge, the tail rotor drive shaft is equipped with two "face gears "that carry the torque across the hinge point for normal operation but disengage when the tail boom is folded. When the fast right hover turn was stopped with full left pedal, the torque in the shaft was so high that the teeth on one face gear were sheared off and the tail rotor stopped turning. Fortunately, the helicopter was just a few feet above the ground so the emergency landing was successful.

What caused the high torque? It was an aerodynamic phenomenon that all rotors can encounter when called upon to quickly increase thrust. This is shown by the results of a whirl tower test in which the collective pitch was increased from zero to 12 degrees at various rates.

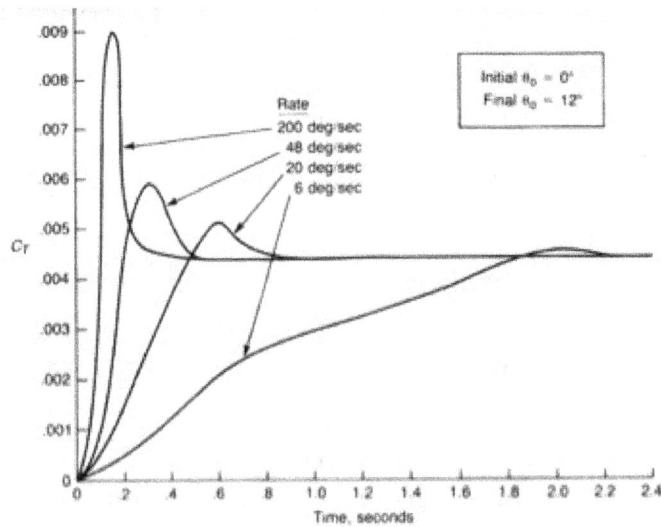

FIGURE 2. **Effect of Rapid Pitch Change on Rotor Thrust**

It can be seen that a very fast increase of 200 degrees per second results is a significant overshoot of thrust compared to the final value. This is because the angle of attack was 12 degrees before the induced flow decreased it. In this case, the flow took nearly a second to be established.

In a fast right hover turn, the main rotor torque is doing most of the work so the tail rotor collective pitch is low. The maximum collective pitch for full left pedal on tail rotors is up to 35 degrees, much more than the 12 degrees of the whirl tower test, so the thrust overshoot would be expected to be somewhat more in the maneuver that we are considering.

Not too much more, however. The rotor, of course, is stalled so that the lift coefficients on the blade elements are limited, but the drag coefficients keep increasing in this angle of attack region as shown for a NACA 0012 airfoil in Figure 28-3. It is the high drag that causes the problem

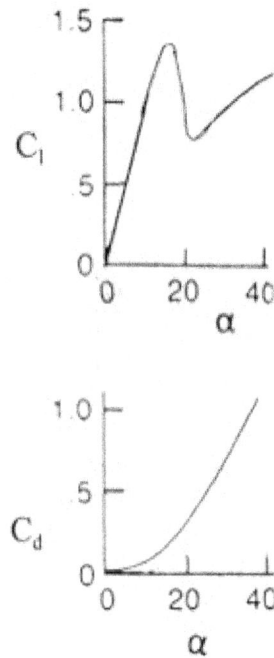

FIGURE 3. Lift and Drag Coefficients at High Angles of Attack

I have calculated the tail rotor torque for my example helicopter assuming that the pilot applied full left pedal and immediately got a collective pitch of 32 degrees and thus a transient angle of attack of 32 degrees. The resultant calculated torque is 9800 foot-pounds. This is compared to a value of 8150 foot pounds that may have been the designer's goal based on 1.5 times the calculated torque for right sideward flight at 40 knots at the helicopter's maximum gross weight.

It is, of course, impossible for the pilot to achieve full left pedal instantaneously so the theoretical torque will not really be generated, but something close to it caused the HSS-2 accident. After realizing the scenario, Sikorsky added dampers to the pedals to make it difficult to get high rates, and all subsequent Sikorsky designs have used them.

I have been told of other helicopters in which the tail rotor shaft got a permanent twist due to this maneuver. Hopefully these incidents can be prevented by telling the pilots, " Please don't do that "

A Possible Tail Rotor Problem

CHAPTER 29   *Can You Fly a Real Helicopter Like a Model?*

Many of us have been amazed while watching what a skilled radio-controlled model helicopter flyer can do by making maneuvers that we have never seen a full-scale helicopter do. What is the difference? And what do we have to do to approach the same results?

Fuselage Passing Rate

First of all, we have to recognize some basic facts of physics. One is that when watching a full-scale helicopter doing a maneuver at 80 knots, its forty-foot long fuselage is passing by at about 3 lengths per second while with a 1/10 scale model flying at 20 knots, the fuselage passing rate is about 8. This makes an observer sense a faster action.

Control sensitivity

Another effect is that the model is much more sensitive to cyclic pitch. I investigated this by comparing the example helicopter of my textbook with its one/tenth scale model version. For this calculation, I have made two non-dimensional factors the same for both aircraft. The first is the Lock Number which is the ratio of blade aerodynamic parameters to blade flapping inertia. The second is the blade loading coefficient, $C_T/2$. I have also used the same rotor tip speed for both.

The results of these calculations are that one degree of longitudinal cyclic pitch produces a pitch rate of about ten degrees per second for the full-scale helicopter, but almost seventy degrees per second for the model! This high response rate would make the model very hard to fly. This is usually taken care of with a Hiller-type stabilization system or a black box with pitch and roll gyros. By adjusting the gains of these devices, a control response can be obtained to fit the skill of the flyer. The higher, the more exciting the air show.

The same tailoring of control power can be achieved on the full scale helicopter. A concern, however, is that at a certain high level of response, a pilot could over-control and end up with an accident due to Pilot Induced Oscillation.

So, You Want to Fly Upside Down

If a radio-controlled helicopter model can fly inverted, so can a full-scale helicopter. The most important modification would be in extending the negative collective pitch range. I have used my example helicopter to determine how much. In the hover performance program, inverted flight was simulated by eliminating the 4.2% fuselage vertical drag and by changing the blade twist from negative ten degrees to positive ten degrees. In this situation, the outer portions of the blades are producing more lift than normal. Where 17 degrees of collective is needed for normal hovering, only three degrees of negative pitch is required up-side-down.

Negative twist is used to improve hover performance, so positive twist decreases it. This is indicated by the Figure of Merit being reduced from 0.73 in normal hover to 0.68 inverted. This loss of performance is partially compensated for by the absence of any fuselage vertical drag.

The only other factors to be considered are those that designers of stunt airplanes have incorporated such as good body restraints, leak-proof tanks, and power plants that run in any orientation. Perhaps some day we will see big helicopters flying like little ones.

CHAPTER 30   *An Attempt to Explain a Difference in Vibration*

The Sikorsky Black Hawk carries 320 pounds of vibration absorbers. The Hughes (now Boeing) Apache carries none. The rotors look about the same, so why the difference?

There are more mysteries associated with vibration than with any other helicopter subject, but there are two configuration differences between the two aircraft that might be important in answering the question. They are blade twist and mast support.

Twist

Blade twist is helpful in improving hover performance. The Apache has - 9o degrees and the Black Hawk has -18 o. That extra -9o improves the hover-out-of-ground- effect performance of the Black Hawk by about 200 pounds.

The high twist, however may be the reason the Black Hawk prototype shook so hard at high speed and motivated the installation of vibration absorbers both in the hub and at several locations in the fuselage.

One of the effects of high twist can be illustrated by the angle of attack distribution on the advancing blade at high forward speeds. The figure compares this distribution for both helicopters at 145 knots at their design gross weights at sea level. It can be seen that both have significant areas of negative angles of attack that are producing downloads on the advancing tip due to the twist that was put in to help them hover. (It is axiomatic that in helicopters, "Whatever helps hover hurts forward flight and vice versa ") Although the two hatched areas are about the same, the Black Hawk's region contains higher negative angles of attack because of higher twist.

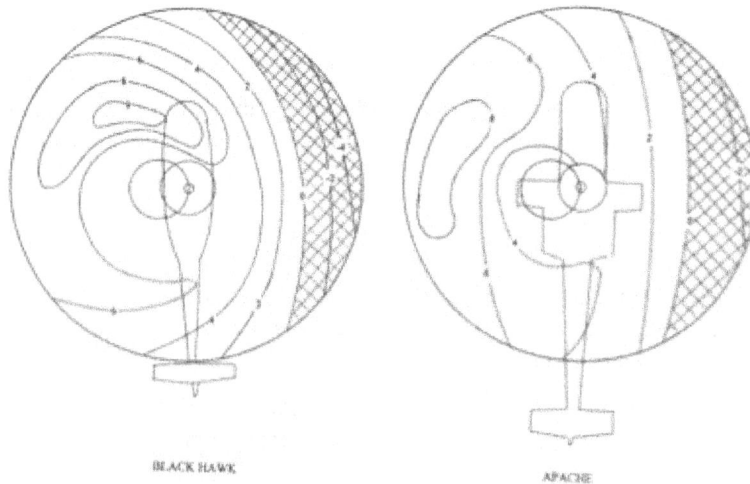

BLACK HAWK

APACHE

FIGURE 1. Angle of Attack Distributions at 145 Knots

A tip vortex is shed wherever the lift goes to zero. The usual place for this is at the extreme tip, but if the lift goes to zero as on the inboard boundary of the cross-hatched region, a vortex spinning in the conventional way is generated while at the extreme tip there will be another spinning in the opposite direction.

Thus the cross-hatched region is generating a swath of vorticity containing some filaments spinning in one direction while others are turning in the opposite way depending on which edge of the region generated them. Some filaments will get together and cancel each other out, but others will get together and produce stronger circulation in one direction or the other. Thus the swath must have a chaotic vorticity characteristic and it is just waiting to change the angle of attack distribution on the next blade coming by. Because of the change in angle of attack distribution on this blade and the resulting blade bending, it will leave a swath different than the one it encountered and this will be true for every subsequent blade passage. No wonder the Black Hawk shook during the first forward flight tests!

Mounting the Rotor

The other difference between the Black Hawk and the Apache is how the rotor is mounted to the airframe. The Black Hawk, like most helicopters, uses a rotating shaft coming up from the transmission with the hub fixed to it. The Apache, on the other hand, uses a "static mast "which is a non-rotating hollow tube containing a quill shaft that turns the hub which is mounted with bearings at the top. Thus the static mast carries bending moments as steady loads rather than as vibration-causing oscillating loads.

One advantage of this system is that it gives the structural designer the flexibility to adjust the longitudinal and lateral natural frequencies of the mast-support system separately to be able to decouple the fuselage response from the rotor inputs. This must have an effect on vibration.

A Little Background

Despite the lack of vibration absorbers now, the Apache started out with a set of pendulum absorbers at each blade root. This was because they had been found necessary on the four-bladed Hughes OH-6 and four blades had been chosen for the Advanced Attack Helicopter (AAH) proposal that became the Apache. The absorbers were on the prototype during its early flight tests, but one day one of the support brackets failed by fatigue and that pendulum was lost. Fight tests other than those associated with vibration were critical at the time, so the other three absorbers were removed while design changes to their brackets were started. After the first several flights without absorbers, the pilot reported that he could feel no difference in vibration from the previous flights and so the absorbers were never put back on. At the Army's suggestion, however, a structural member was installed in the front of the cockpit on which a spring-mounted weight could be mounted as an absorber, but it was never found necessary.

Lady Luck may Play a Part

After Apache production was well under way, Hughes decided to join the new trend into composite blades. New blades were designed with the stipulation that their weight and stiffness distributions, natural frequencies, and outside contours should be the same as the metal blades.

In high speed flight, much to our surprise, the test Apache shook twice as much as with the metal blades! Even after a redesign to make an even better match, the Apache shook. Extensive analysis reported by B.P. Gupta in a 1984 AHS Forum paper could not pinpoint the reason.

His rueful conclusion was, "Serious questions are raised about how well we understand vibration producing mechanisms. In the author's opinion, analytical methods of vibration prediction are inadequate. "Does that still apply today?

CHAPTER 31 *Wishful Thinking*

What We Have Now

One of the big contributors to the science of helicopter aerodynamics was Jan Meijer Drees. His first contribution was made as a young university student in The Netherlands when, in 1950, he produced a movie called "The Flow Through A Helicopter Rotor. "He used a two-bladed model helicopter rotor in a wind tunnel with streams of smoke to show the various paths of air during different test conditions.

From one sequence, at low forward speed, you can see the up-and-down flow at the leading edge of the disc that jumbles up the local angles of attack and accounts for why a helicopter shakes when making the transition from hover to forward flight

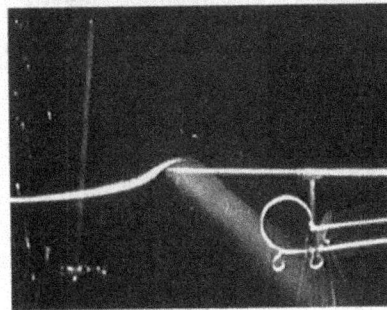

FIGURE 1. Transition Flight

In another sequence, at a wind tunnel condition representing near vertical descent, the reingestion of the wake that is reducing the blade angles of attack helps to explain why pilots report, "It dropped right out from under me "in the "settling with power "or "vortex ring "state.

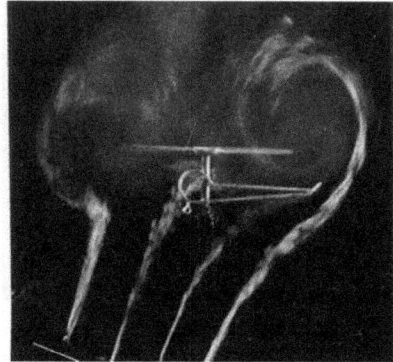

FIGURE 2. Air being Re ingested in the Vortex Ring State

(The smoke streams were made by running electrical heating wires through capsules filled with oil-soaked cotton.)

After he graduated in 1952, Jan and several other engineers formed a company that designed and produced (though only 25) a small ram-jet powered helicopter called the "Kolibrie "(meaning "Hummingbird', a name--naturally- for several other European helicopters). It was used for agricultural work in both The Netherlands and in Israel as late as 1981.

Jan came to America to work for Bell in 1959. He was instrumental in many Bell programs and retired as Vice President of Technology.

But How About Now?

To get to my point: It's now been more than sixty years since Jan made that movie. There has been a slight improvement in technology since then but it has not been used to make another one.

I can visualize a new wind tunnel project using a model of a modern rotor and filming the results with several surrounding color cameras with digitally generated grids. Instead of using smoke, I suggest simultaneously dropping ten pieces of different colored confetti pieces from the same spot for each run to capture the amount of chaos.

Another Suggestion

That's not the end of my dreams. We can use computer programs of various degrees of sophistication to calculate the angle of attack at each blade element in various flight conditions, but accurately measuring them in a wind tunnel or in flight has not yet been possible. The closest attempt I know is a round-about method involving comparison of measured pressure distributions with those measured in a two-dimensional wind tunnel test for the same airfoil at various angles of attack.

Airplane aerodynamicists have a simpler job, for instance, in measuring the angle of attack at the horizontal stabilizer as affected by the reasonably distant wing-tip vortices. A simple mechanical vane will do.

Attempting to do the same on a rotor blade is difficult because of the high centrifugal forces that might lock up the vane's bearings. And also our tip vortices are not reasonably distant from the blade element and so changes in angle of attack might be too fast for a mechanical vane to respond to.

My suggestion is to mount a camera at the root of the blade. To constrain the picture to the desired blade element, a light source above the blade on an arm mounted on an upward extension of the rotor shaft would make a narrow plane of light at the right location. Again I would use a few different colored confetti pieces and a shutter speed just long enough to get colored lines to represent stream lines. To define the blade pitch, the blade would have a "nose spike "at least a chord long.

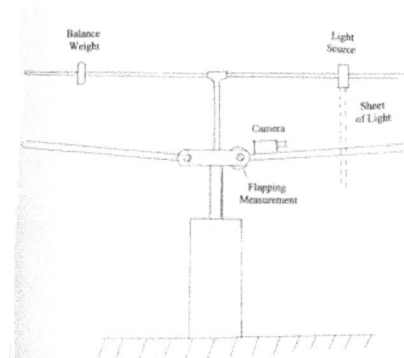

FIGURE 3. Set-up for Measuring Airflow at Blade Element

The tunnel log book for each run number would list the blade station, the tunnel speed, the shaft angle, the collective and cyclic pitch settings, and the tunnel measurements of lift and drag. Each photograph would include the run number, the azimuth angle, and the flapping angle.

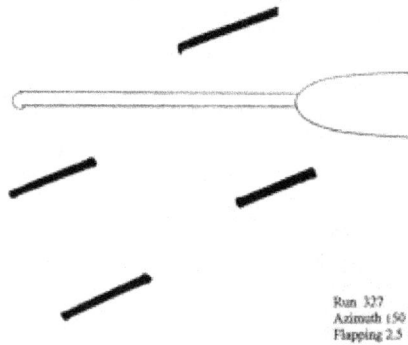

Run 327
Azimuth 150
Flapping 2.5

FIGURE 4. **What the Camera Would Catch at the Blade Element**

I can dream, can't I?

CHAPTER 32   *The Lock Number*

In the late 1920s, a group of British aerodynamicists were deriving equations for that new aircraft, the Autogyro. One of them, C.N.H. Lock, found that a combination of parameters came up several times in his equations. The Lock number, as we helicopter people call it in honor of that pioneer, represents the ratio of aerodynamic to inertia forces on a blade and is widely used in our work. For blades deigned for modern helicopters, its value falls into a relatively small range from about 4 to about 15 no matter how big the helicopter is.

I do not usually put equations into my columns, but this will be an exception. The equation for the Lock number is:

$$\gamma = \frac{c \times \rho \times a \times R^4}{I_b}$$

(Note that we have assigned a Greek letter to it as we often do in aerodynamic equations. Unfortunately, this letter, γ ?, is also used in other places; for climb angle, for a rate parameter in dynamic stall analysis, and for several others. The reason for this problem is that the Greek alphabet has only 24 letters--not enough to go around in aerodynamic equations. (One of my Chinese friends says, "That's no problem for me, I have 5000 characters ")

The parameters in the numerator are:

- c = chord, in feet
- ρ = density of air, in slugs per cubic foot
- a = slope of the lift coefficient per radian angle of attack =5.73, non-dimensional

R= blade length outside the flapping hinge, in feet

The parameter in the denominator is:

- I_b = moment of inertia of the rotor blade

By combining the units, it can be shown that the numerator is in slug-feet-squared. This is also the same units for the blade Moment of Inertia, outside the flapping hinge. in the denominator. Thus the Lock number is a non-dimensional parameter.

Why the Difference?

The Robinson R-22 blade has a Lock number of 4.19. This low number represents a relatively heavy blade compared to other helicopters where an average value is about 8. Small, single-engine helicopters such as the R-22 have a special problem concerning the rapid loss or rotor RPM in case of an engine failure. Pilots of early R-22s complained about this and tip weights were added to improve the situation, thus lowering the Lock number.

On the other hand, for the Sikorsky CH-53E with three widely spaced engines and at least three fuel tanks, an instantaneous complete loss of power is very unlikely and so relatively light blades producing a Lock number of 14.87can be safely used.

Getting the Moment of Inertia

An aerodynamicist calculating the Lock number has no trouble with the numerator, but determining the Moment of Inertia may be a challenge for him. He can get it by adding up the masses of each blade element multiplied by the square of its moment arm from the flapping hinge. He should be careful to account for an uneven mass distribution such as one due to a concentrated weight.

Or he can use an approximate calculation by considering the blade as an uniform rod. In this case, the equation is:

- $I_b = \dfrac{1}{3} \times \text{Total Mass} \times \text{Length}^2$

A more accurate procedure is to treat an existing blade as a pendulum. This requires the time, t, in seconds, to make a single oscillation and the distance in feet from the support to the center of gravity, x_0, as used in the following equation:

- I_b = Total Mass $\times x_0 \times t^2$

FIGURE 1. Set-up for Experimentally Determining Moment of Inertia

Or yet another method is to go across the hall to the Dynamics Department and ask. "What is the Moment of Inertia of this blade?'

What the Lock Number Effects

Probably the first calculation that uses the Lock number is for the coning in hover. It is determined by setting the sum of the moments at the flapping hinge due to lift forces, centrifugal forces, and the blade weight to zero. This results in the equation:

$$I_b = \frac{2}{3} \times \frac{\gamma \times \dfrac{C_T}{\sigma}}{a} - \frac{1.5 \times g \times R}{(\Omega \times R)^2}$$

The first term is due to both the lift and centrifugal forces and the last to the weight of the blade.

The example helicopter in my textbook hovers at sea level Blade Loading Coefficient, $C_T/2$, of 0.086 and its Lock number is 8.1. The equation gives a coning in degrees of 4.3o. If the last term, due to blade weight is omitted, the calculated coning increases to 4.4o. So we can say that lift and

Lock number are important, but that the weight of the blade has a very small effect on coning. The equation in forward flight gives essentially the same coning.

Another application of the Lock number is used to determine how fast the blade responds to a control input. All of the cockpit controls are rate controls. This is especially evident in hover. A step input in one will eventually result in a steady rate. A measure of the speed of the response is called the "time constant. "Since some results take a long time, the time constant is based on the time required to obtain only 63% of the final rate. (You need to get a dynamicist to explain where the 63% comes from.) For a helicopter with a teetering rotor, the time constant equation is:

$$T_{63} = \frac{16}{\gamma \times \Omega}, \text{ seconds}$$

With offset flapping hinges, the resulting equation is somewhat more complicated, but gives nearly the same result. For my example helicopter, the time constant is 0.09 seconds during which time the blade travels only 1300 in azimuth--about a third of a revolution. Even for helicopters with different Lock numbers the response to control will be quick. This is the reason that once a pilot learns to fly a Robinson R-22, he will not have much trouble flying a Sikorsky CH-53E.

Without showing the equations, I want you to know that the longitudinal and lateral flapping while the helicopter is pitching or rolling are also affected by the Lock number. This determines how much cyclic pitch is needed to hold the helicopter steady in a maneuver.

The Lock number also enters into the equation for how much hub moment is produced by flapping in a rotor with hinge offset.

Several other aspects of helicopter flying are affected by Mr. Lock's number, but this seems to be a good place to stop.

CHAPTER 33   *The Cyclogyro*

Besides the helicopter and the autogyro, there is yet another rotary wing aircraft that has intrigued inventors both in the past and in the present. It is the Cyclogyro, or "Paddle Wheel Aircraft. "The rotating wing on this concept, unlike the others, does not form a disc but a cylinder as shown in Figure 33-1. In forward flight, each blade traces out a cycloid in the side view, thus accounting for the name.

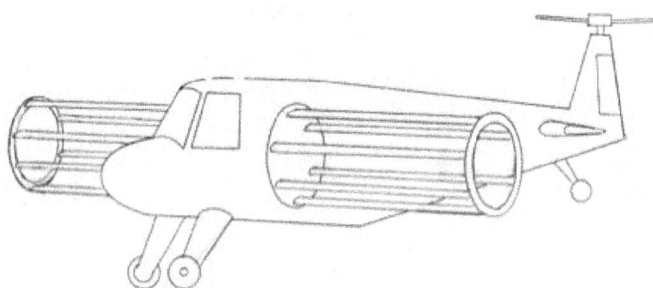

FIGURE 1. **A Possible Cyclogyro Configuration**

Compared to the conventional rotor, it has the advantage that each blade element is operating at the same velocity and thus at its full capacity. It has the disadvantage, however, that the centrifugal and aerodynamic loads do not compensate for each other as they do on a rotor with flapping hinges.

In the Beginning

The history of this configuration can be traced to vertical-axis windmills in ancient China and Persia. The first US patent for a "feathering vane windmill "was issued in 1824 and the concept has been recently revived as the Darius windmill.

As a propulsive device, it was first used for marine propulsion as the "Paragon Propeller "in England in the early 1900s. Since it can serve both as a propeller and a rudder, it now is used on many marine craft that require good maneuverability; such as tugs, ferries, and river boats.

It was early recognized as a system that could be used on aircraft. Several attempts were made during the 1920s and 30s. Three are shown in Shell Oil's 1952 movie, The History of the Helicopter. But, alas, I can find no photo or film clip showing light under the wheels! Tests were also done by NACA in their full-scale wind tunnel.

One of the pioneers who worked out the theory of the device was Professor Kurt Kirsten at the University of Washington. Initially his interest was in explaining the flight of birds whose wing tips trace out cycloidal patterns. He then converted this understanding to developing mechanical devices. Starting in 1921, he made wind tunnel tests of small models and then in 1923, with the help of the Boeing Company, he tested the 15 foot diameter, 400 horsepower rotor - shown in Figure 33-2 - which was suitable for a Cyclo-copter. (In his report Kirsten says, "The test of the second model took place at a time when aeronautical engineering was suffering from the effects of a general helicopter epidemic ") The set-up as shown in the photo was arranged to produce a horizontal force, and we are looking at the downwind side.

FIGURE 2. **Professor Kirsten's Cyclogyro Rotor**

The tests were plagued with fatigue failures of that new light-weight metal, "duralumin "which could have been avoided if the benefits of alloying and heat-treating aluminum had been known at the time. Following this disappointment, Kirsten transferred his attention to marine propulsion and most of these present-day systems can be traced to his work.

A more recent (2005) aeronautical application was on the Cyclo-Crane which surrounded a spinning helium-filled blimp with a set of cycloidal vanes for help in lifting very large loads. Work on cycloidal propulsion in air has continued in recent years at several universities and research labs resulting in impressive performance in terms of pounds of hover thrust per horsepower.

A Difference

Cycloidal propulsion in air has different requirements than in water. On a boat, the device is used primarily as a propeller producing a force parallel to the course and only occasionally as a rudder to produce a side force. In air, on the other hand, the force vector of the device is primarily supporting the weight of the aircraft. It is straight up in hover and nearly perpendicular to the flight path in forward flight where only a small tilt of the force vector will be sufficient to propel the aircraft forward. For example, if the aircraft weight-to-drag ratio is 10, then the total thrust vector only has to be tilted forward by about 15 degrees.

How it Works

In hover and low speed flight, the blade pitch can be made to change cyclically with a relatively simple "four-bar linkage "system. Figure 33-3 shows how this can result in a vertical force in hover that can be used to support the aircraft or with a change in phasing, produce a tilted force that can both support the aircraft and propel it through the air. Most of the upward thrust is developed when the blades are at the top and bottom of the circle. (I have included a "top spot "on the "upper "surface of the blades to show that they turn over during the cycle and therefore a symmetrical airfoil is required.). The magnitude of the vector could be controlled either with a "collective pitch "input or a change in RPM. Unfortunately, the pitch schedule that is good for hover and low-speed flight is not so good for fast forward flight.

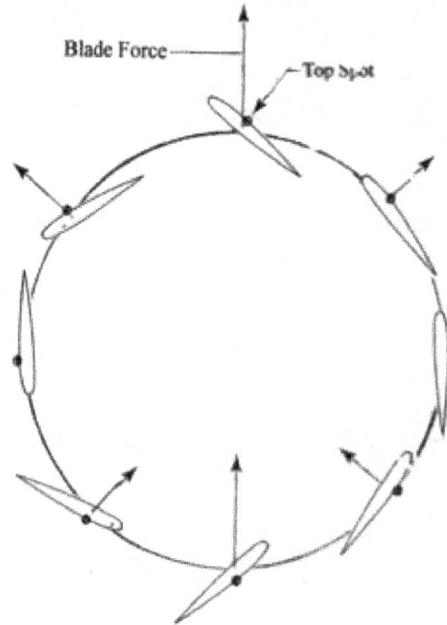

FIGURE 3. Cyclogyro Rotor in Hover

Going Fast

For fast forward flight, conditions are different. Now we have to consider an "advance ratio "similar to the helicopter's "tip speed ratio. "It is zero in hover, of course, but has a finite value as the cyclogyro picks up speed. At first, the blades will be tracing out cycloids with loops, just as a helicopter's tips in a top view trace out similar pattern. At some forward speed, with the blade schedule suitable for hover, some of the blades will be in the "reverse-flow region "as dictated by adding the forward and the rotational speed vectors. These blades will not be producing optimum aerodynamics.

A special Case

If the forward speed is high enough to match the rotational speed, the blade speed ratio will be unity and the cycloids will have no loops. This represents the path of a pebble caught in the tread of a car's tire. This would correspond to a helicopter with a tip speed ratio of one. For this condition, the blade pitch schedule should be as shown in Figure 33-4 which is different than in hover.

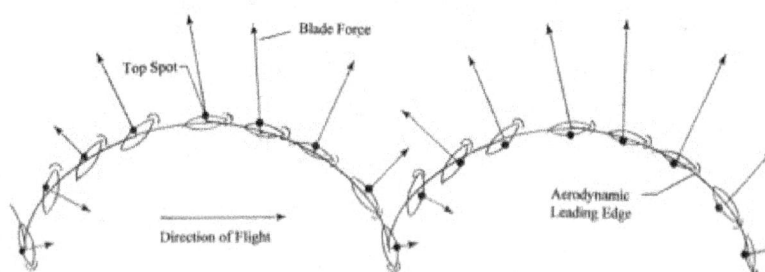

FIGURE 4. Cyclogyro Path at an Advance Ratio of One

Note that the blades turn over only once in two cycles, and that the aerodynamic leading edge moves from one side of the blade to the other. This requires that the airfoils be double-ended. This schedule can be achieved relatively simply with gears as Kirsten did. This is the schedule that he used even in his Cyclo-copter experiments as can be seen from the photo, but it is not the best way to get good hover performance.

Of course, any other forward speed between hover and high speed optimum performance requires its own special feathering schedule which cannot be achieved with the mechanical systems that work for hover or for high speed forward flight near an advance ratio of one. For off-design speeds, other control laws must be used. The logical system is one using

"Independent Blade Control "(IBC) where each blade has its own actuator which is controlled by a signal from a computer that has been programmed to select the best blade angle for the flight condition. (To see how IBC with its advantages and disadvantages is being proposed for helicopters, go to the Spring, 2004 issue of Vertiflite, or Chapter 85 Helicopter Aerodynamics, Volume II).

A consideration, however, is while IBC as proposed for helicopters controls to make only a small pitch change and thus is only a small addition to the basic control system, in a cyclogyro it would be the basic control system and thus requires 100% reliability.

A Final Complication:

Just as on a helicopter, the torque associated with the power required to operate the cyclogyro rotors will tend to rotate the fuselage in the opposite direction. This will require something like a horizontal tail rotor of Figure 33-1 which then could also be used for pitch control. Another option to solve the torque unbalance is to use a tandem-rotor configuration with the two pairs of rotors operating in opposite directions. Roll control will require differential collective and yaw control needs differential thrust tilt (or a tail rotor). These control challenges make the IBC system look like the only way to go.

CHAPTER 34   *The Nano Hummingbird*

FIGURE 1. The Nano Hummingbird

We don't often see an aircraft flying during the presentation of a paper at an AHS Specialists "Conference, but those who attended the one in San Francisco in January of 2012 did. Matt Keennon of the AeroVironment company flew his battery-powered Hummingbird all around the conference room. Technically, this is an "Ornithopter "

The company, founded by Paul MacCready of man-powered airplane fame, has been busy in developing un-manned aircraft for the military for many years. In 2005, DARPA opened a competition for a "Nano Air Vehicle "(NAV) in the shape of a hummingbird that could carry a color television camera the size of a pea into an open window or door to see what was in there.

AeroVironment won the contract in 2006 and assigned Keennon and an experienced team to the project. Four and a half years and $4 million later the "Hummingbird Drone "was a cover subject for Time magazine and I was invited to the plant for a personal demonstration.

The development had gone through fourteen prototypes involving approximately 300 investigations of wing shapes and materials and many arrangements of the flapping mechanism which rotates and twists the wings. Much of the work was done under a microscope.

The result is a tiny aircraft with only two flapping wings and no tail surfaces that can climb, pitch, roll, and yaw in response to the remote pilot's wishes. It has a wing span of 6.5 inches and the gross weight is less than a AA battery. The wings beat thirty times per second, and it can hover and fly forward at eleven miles per hour. It has an endurance of eleven minutes before the battery needs recharging.

Since it is inherently unstable, the control system includes miniature gyros and actuators to stabilize it. It can be controlled by a pilot using a flight controller who has it in plain sight or from what he sees "heads-down "on his monitor from the TV camera.

During my visit, I saw it fly and was also shown a PowerPoint presentation of the project. One slide that got my attention showed that the maximum Figure of Merit was only 0.2. This low value, compared to a higher value for a rotor-supported aircraft, is due to two sources. The first is the high drag coefficients associated with the very low Reynolds numbers that exist. The second, compared to a rotor where the blade elements are producing continuous lift, is that on a flapping wing during part of the stroke, the wing is not producing lift, but still has drag.

While real hummingbirds are essentially silent while hovering, the Nano Hummingbird sounds like a giant Beetle. This detracts from DARPA's stealth desire. Feathers are needed.

CHAPTER 35   *The Other Sikorsky X-2*

How it Started

Back in 1951, the Army and the Air Force got together and put out a Request For Proposal for a helicopter that would fly faster than normal. Three companies were given contracts for what, at that time, were known as "Convertiplanes. "The name applies because they were helicopters that were converted into airplanes at high speeds.

The Three Winners

The first winner, by designation, was McDonnell's XV-1. This was a winged autogyro with a pressure-jet rotor used for hovering but which was unloaded and slowed down for high-speed flight. A piston engine could either drive an air compressor or a propeller. Two prototypes were built and successfully flown, but the aircraft was not put into production.

FIGURE 1. **The McDonnell XV-1 Convertiplane**

The second winning design was Sikorsky's XV-2 (or S-57). This aircraft used a counter-bal-anced, single-bladed pressure-jet rotor for hover and low-speed flight. Power was to be supplied by a jet engine whose exhaust could either be used to drive an air compressor to feed air to the rotor or to propel the aircraft as a jet airplane. The rotor was to be unloaded and stopped when the speed was high enough for the wing to provide enough lift to support the gross weight. The rotor was then to be stowed in the tail boom and the jet exhaust directed aft.

FIGURE 2. Artist's Version of the Sikorsky XV-2 as it Might have Been

FIGURE 3. Artist's Version of the Sikorsky XV-2

The third Convertiplane was Bell's first Tilt Rotor, the XV-3. Both rotors were driven by a piston engine in the fuselage. Two prototypes were built and successfully flown. Although the XV-3 was not put into production, the configuration is now flying as the Marine's Osprey.

FIGURE 4. **The Bell XV-3**

Sikorsky Problems

While the other Convertiplanes flew, the Sikorsky design did not get any further than wind tunnel tests.

Two wing designs were tested on the wind-tunnel model. One was a "delta "which would have given future configurations supersonic capabilities. The other was a more conventional tapered wing which would have been used on a prototype.

Sikorsky had two problems at the time. Their facilities were tied up supporting full production for the Korean War and during the 1950s there was no suitable jet engine available for the XV-2. Therefore, in 1960, the project was canceled.

On Second Thought

That was not quite the end of the XV-2, however. In 1964, somebody at Wright Field became interested in the design and asked the question, "If you are in airplane flight and you lose the engine, can you restart the rotor to make an autorotative landing?'

They gave Sikorsky a contract to put the model back in the United Aircraft Corporation wind tunnel in Hartford, Connecticut to answer the question. I had recently joined Sikorsky with a bit of wind tunnel experience and so I was assigned to this project.

To represent normal rotor-flight conditions, the rotor was powered with shop air. Stopping the turning rotor in the tunnel was done by shutting off the air and using the drag on the blade elements--and in the final stages--a brake. Of course, when either starting or stopping, the flapping hinge had to be locked to avoid excessive flapping that could have cut the tail-boom off.

The question was, "In airplane mode can the rotor be started without air "When starting without air--representing a power failure--it had to use aerodynamic lift on the blade. This was accomplished by pitching the model up and using a little negative blade pitch. The air bouncing off the bottom of the stationary blade and the "Coanda "effect of the flow around the nose of the airfoil started the rotor in the desired direction. From then on, it was in autorotation and the answer to the question, as demonstrated in the wind tunnel was "Yes "

FIGURE 5. The Wind-Tunnel Model, Blade on right, Counter-Weight on Left

Yes, But?

The fact that it was a one-bladed rigid rotor turning slowly with lift during the first several revolutions meant that it produced a series of highly transient rolling and pitching moments. Like other wind tunnels of the time, the mechanical balance system could not keep up with the changing moments so the response during the first revolutions was calculated by using measured moments with the rotor stopped every15 degrees of azimuth. The conclusion was that it would have given the pilot an exciting ride.

On Third Thought

For this reason, Wright Field gave Sikorsky another contract to design a two-bladed stowable rotor and test it in the wind tunnel. A person could represent the stowing mechanism by kneeling on the floor with arms out-stretched and then bending forward while folding the arms back.

I had left Sikorsky to go back to school when this configuration was tested in the wind tunnel, but I have been told that the tests were successful and had the project been revived, this is the design that would have been used.

36

CHAPTER 36   *A Revived Configuration*

If you are a newcomer to the world of helicopters, you might think that the high-speed Eurocopter X3 is a new idea. If you are an old-timer, you might remember a similar configuration from long ago.

FIGURE 1. The Eurocopter (now Airbus)X3

A Long Time Ago

Fairey Aviation in England began building helicopters right after World War II with the not-too-successful "Ultra-Light. "The next project was the three-bladed, 4400 pound "FB-1 Gyrodyne "with a 525 horsepower piston engine. This was a more-or-less conventional helicopter except that instead of a tail rotor to provide for anti-torque and directional control, it used a single propeller at the tip of a stub wing on the right-hand side.

With this configuration, the rotor did not have to provide the forward thrust as on a conventional helicopter. This eliminated the possibility of retreating blade stall and in 1948, the aircraft set an official speed record of 124 miles per hour!

FIGURE 2. **The FB-1 Gyroplane**

A Redesign

The British Army was all ready to give Fairey a production contract when the prototype crashed killing the two crew members. The Army then decided to buy Sikorsky S-51s instead. But this was not the end of the Gyrodyne. Fairey redesigned it with a two-bladed pressure jet rotor. This type of rotor had been developed by Doblhoff in Austria during the war. It used air from a compressor in the fuselage that was run out to nozzles in the blade tips where kerosene was burned to produce thrust.

The engine of the new Gyrodyne also drove propellers on both wing tips, this time as "pushers. "This aircraft is referred to as the "Jet Gyrodyne. "It's first flight was in 1954.

FIGURE 3. **The Fairey Jet Gyrodyne**

As with all helicopters with tip-driven rotors, the fuel consumption was high. Without auxiliary fuel tanks, flights were limited to about 15 minutes. But the purpose of this aircraft was to produce knowledge for a later one, the "Fairey Rotordyne "

Going Big

The 40-passenger Rotordyne was meant to be a "flying bus "serving as a transport from city center to city center. Like the Jet Gyrodyne, it had a pressure jet rotor which was intended to be used only for take-offs and landings, so high fuel consumption was not a problem. Most of the flight was as an autogyro at 160 knots with the wing carrying about half the gross weight. The engines on each wing drove propellers or air compressors when required for low-speed operation.

FIGURE 4. The Fairey Rotordyne

Marketing found interest all over the world. Kaman Helicopters was planning to build it in the United States under license. The beginning of construction was to start based on forty firm orders. They never came.

Too Noisy

The main reason was that while burning fuel in the rotor tips, it was extremely noisy. One measurement of sound level from a microphone 600 feet from the take-off spot was 115 db which is enough to do ear damage and to make it a very unwelcome guest to any city center. When demonstrating at air shows, the take-off of the Rotordyne was done only when a squadron of jet fighters was making a high-speed pass in front of the audience. Several attempts were made to reduce the noise, but it was too late. Fairey had been absorbed by Westland and the British government was cutting back on helicopter developments.

CHAPTER 37   *Jack Real - His Two Helicopter Careers*

Like many others, Jack Real (1915-2005) entered the world of helicopters through an unantici-pated sequence of events, but once he was in, his contributions were very significant.

FIGURE 1. Jack Real

With the end of World War II, aircraft companies were left with large engineering staffs and manufacturing plants that had little to do. Each company looked for alternate products to keep their assets busy. Lockheed looked at the light airplane field and built several prototypes, but didn't pursue this possibility. Instead, Robert Gross, the Chairman of the Board of Directors, became interested in helicopters. This was started when Irven Culver, Lockheed's resident genius, expressed some unique ideas (for the time) concerning hingeless rotors based on his observations of the stable flight of a rotating toy.

Irv was given the assignment of putting his ideas into practice by building the world's first radio-controlled model helicopter. The model, with blades cantilevered from the hub without hinges and with cyclic pitch controlled by a gyro exhibited such good flying qualities that in July of 1959 a small "Skunk Works "was established in a walled-off corner of the experimental hangar in Burbank, California where five engineers' of which only one had previous helicopter experience-and a small group of mechanics designed and built the two-place, piston-powered CL 475 (the 475th preliminary design project of the California Lockheed company).

FIGURE 2. The Lockheed CL 475

After completion, the aircraft was put under wraps and trucked to a remote corner of the Mojave Desert for flight testing. After overcoming the same type of development problems that had been encountered by other pioneers, the CL 475 became a good enough flying machine that military and FAA pilots were invited to fly it. They made glowing reports about its exceptional flying qualities.

Jack Real's First Helicopter Career

The project gradually lost its Skunk Works character and the regular engineering and flight test departments became active with the help of some newly-hired people with helicopter experience (including me). Thus in his twenty-first year at Lockheed, Jack Real got into the helicopter business since he was Manager of the Lockheed Flight Test Department. To help prepare for this new career, he took my after-hours helicopter aerodynamics course.

Lockheed bid on the LOH (Light Observation Helicopter) project using the "rigid rotor "design developed for the CL 475, but was not chosen. However, as a consolation prize, the

Army and the Navy combined to give Lockheed a research contract for a somewhat larger helicopter that became the XH-51A of which several were built.

FIGURE 3. The XH 51A

Jack Real was to have an active role in its test and in the FAA certification of the civilian version, the Model 286. Two of these were built, but Lockheed decided not to put them into production. Instead they were sold to an airplane collector. They were destroyed when his hangar burned down.

FIGURE 4. The FAA-Certified Lockheed Model 286

Jack was also involved in the testing of the compound configuration of the XH-51A with a wing and jet engine. It reached a level speed of 263 knots.

FIGURE 5. The XH-51A Compound

This was part of the Army's research program that led to the competition for the 220-knot AAFSS (Advanced Aerial Fire Support System) proposal which Lockheed won in 1965 with the AH-56A Cheyenne. At the same time, Jack became Lockheed's Vice President for helicopters.

FIGURE 6. The Lockheed Cheyenne Anti-Tank Compound Helicopter

During the next six years, Jack worked on the Cheyenne "12 hours a day and six days a week "while ten pre-production aircraft were built and tested during a fairly successful engineering program. It was not enough, however, to overcome the mounting political, financial, and change-of-mission considerations that were to convince the Army that what they really wanted was a somewhat slower and simpler aircraft--which they got in the Apache. When Jack later talked

about the final Cheyenne days, he came to a point where he would say, "It's about here that I start crying "

Another Life

During this time, Jack was leading another life that consumed much of his spare time. It had started in 1957 when Lockheed management suggested that he contact Howard Hughes concerning the purchase of some turbo-prop Electras for his airline, TWA. Jack and Howard immediately bonded on the basis of their interest in airplanes. Although TWA never bought any Electras, the personal chemistry lasted the rest of Howard's life. The story can be found in Jack's book, "The Asylum of Howard Hughes "(Xlibris Corporation)

While still working for Lockheed, Jack helped Howard move from California to Las Vegas in 1966 when his state income tax bill became oppressive. Hughes bought the Desert Inn Hotel and became a recluse in it. His near-fatal accident with the XF-11 in 1946 had left him with an addiction to pain-killers and other drugs. This and some unusual health phobias made it impossible for him to lead a normal life.

During the Las Vegas years, Jack helped Hughes buy AirWest Airlines. He also consulted on other matters. One of these had to do with the second competition of the Army's LOH program that Hughes Tool Company Aircraft Division had won the first time with the OH-6A in 1962. At that time, Hughes was so eager to get into the helicopter business that he took a $10,000 loss on each helicopter. For the second production contract in 1968 in competition with Bell with their OH-58, Howard was reluctant to continue that loss and asked Jack to recommend a new asking price. Jack was nearly exhausted from his continuing work on the Cheyenne, but agreed to spend one weekend on this while making Howard promise not to change it. He came up with a price of

$53,400. At the last moment, despite the promise, Howard raised it to $59,700 and lost out to Bell whose bid for the OH-58 was $53,450.

FIGURE 7. The Hughes OH-6

While initially disappointed, Jack later said the loss was probably good. He speculated that the Army would not have chosen Hughes Helicopters to build the Apache if it had still been producing the OH-6A.

There are, of course, many Hughes stories. One that Jack told to a group of us concerned Howard's belief that he should be able to control him. Jack was vacationing at a remote beach where the only line of communication was the telephone in the beach store half a mile from his cabin. One day when Howard called, the messenger from the store could not find Jack who happened to be in the outhouse at the time. When Howard later heard Jack's excuse, he demanded that Jack schedule his outhouse visits to precise times of the day so that he could always be reached.

For Howard Hughes, being one of the richest men in the world after selling TWA was not as pleasant as you or I would think. After four years in Las Vegas, he decided he had to leave the United States because of troubles with his federal income tax. Jack helped move him to the Bahamas where he once again became a recluse in a hotel.

Shortly after that, in 1971, with the Cheyenne program coming to an end, Hughes made Jack "an offer he couldn't refuse. "Howard's intent was for Jack to run all of the aviation assets of his empire including Hughes AirWest Airline and Hughes Helicopters. (Hughes Aircraft was a separate company by this time, working with avionics and missiles.) It turned out, however, that office politics made the arrangement moot. At this time, Hughes, in the Bahamas and in very poor health, was completely isolated by his aides to the extent that Jack could not contact him, nor could Howard contact Jack. Howard was even told that Jack had quit, left his wife, and was living in Europe. This situation lasted for about a year with Jack still on the payroll trying to enlarge the route structure of Hughes AirWest by merger or acquisition, but having nothing to do with the helicopter company. Communication was finally reestablished and Jack was with Howard for the next four years as he moved from the Bahamas to Nicaragua to England and then to Mexico. Jack was on the airplane with Howard from Acapulco trying to get to a hospital in Houston when Hughes died.

Jack Real's Second Helicopter Career

After Howard's death in 1976, Jack remained with the Summa Corporation which controlled the Hughes Estate and was run by Howard's cousin, Will Lummas. Hughes Helicopters had won the contract for the AH-64 Apache in 1973. At that time, Lockheed decided to get out of the helicopter business so I went across town to work on the Apache at Hughes in Culver City. During the next few years, we were busy building and testing Apache prototypes. There were, however, some problems and in 1979, Lummas asked Jack to become the president of Hughes Helicopters. He reluctantly agreed and the next time I saw him was on his first day at the plant when he came into my office to ask for a Table of Organization so he could see who was working for him.

At the time, the Apache problems were so severe that the Army was considering canceling it. The high-mounted T-tail that had been part of the YAH-64 prototype configuration was producing both aerodynamic and dynamic problems that the Army found unacceptable. Jack set an ambitious 30-day schedule for the design, construction and testing of a low-mounted stabilator. The success of this change and his improvement in relationships with the Army as "the customer "turned the program around.

He also made the decision to move the company from the Los Angeles area to Mesa, Arizona to take advantage of safer flight test conditions and with more room to expand. Another of Jack's accomplishments in 1983 was to license the construction of the Model 300C helicopter to Schweizer Aircraft of Elmira, New York.

FIGURE 8. Hughes/Schweizer 300C

Howard Hughes had died without leaving a will, so the Summa Corporation had to come up with $500 million to pay state and federal inheritance taxes. The solution in 1983 was to put the helicopter company up for sale. There were six interested parties, and the winner was McDonnell Douglas.

Jack was asked to stay on as president of McDonnell Douglas Helicopters which he did until he retired in 1988. Not wanting to live an idle retiree's life, he immediately took on a new job as Chairman of the Board of the Evergreen Aviation Museum in McMinnville, Oregon to where he was instrumental in moving the Hughes Flying Boat.

CHAPTER 38 *Wing vs. Rotor*

When an airplane aerodynamicist is suddenly told that he is now a helicopter aerodynamicist, he naturally brings some fixed-wing concepts with him. Some of them cannot be used.

Wing Loading and Disc Loading:

One of the parameters he has been using is "wing loading" in pounds per square feet. When he sees a helicopter parameter named "disc loading" in pounds per square feet, he might assume that they are somehow related. But they aren't. In selecting an engine for a new airplane, two speeds are important, the landing (or stall) speed and the top speed. The landing speed is controlled by wing loading and maximum lift coefficient. Once the desired landing speed and maximum lift coefficient (as achieved with flaps and slats) are defined, the acceptable wing loading for the desired landing speed can be calculated and for an estimated gross weight, the wing area can be found.

An airplane, almost by definition, is a streamlined object. Its drag therefore is primarily due to wing induced drag and the skin friction—referred to as "parasite" drag-- based on a non-dimensional drag coefficient times skin area. This product has the dimension of square feet. It is a factor that the airplane aerodynamicists have not named and do not use in their equations. They prefer to keep the coefficient and the skin area separate.

The helicopter, however, is not a streamlined object (think of the Bell Model 47). The helicopter aerodynamicist uses that airplane factor as "equivalent flat plate area" to account for the many components producing " bluff body" drag. The concept here is to compare the parasitic drag of a component-- or the entire helicopter- with that of a flat plate of a particular size with a drag coefficient of 1. Small helicopters have equivalent flat plate areas of 5 square feet while big ones can be up to 50.

The parasite drag is calculated by multiplying the equivalent flat plate area by the dynamic pressure of the flight speed. The parasitic power- in horsepower- is then found by multiplying that drag by the flight speed in feet per second and dividing by 550. (This constant was invented by James Watt when selling his steam engines to coal miners. It represents what power a "standard horse" can produce, measured in foot-pounds per second.)

Finding how much power is needed for hover is where disc loading comes in. The induced power required to hover can be calculated from that well-known equation, " Force Equals Mass Times Acceleration" which applies directly to the one-time event such as firing a cannon. But it can also be applied to a steady force such as thrust of a rotor to find "induced power." In this case, the mass is the amount of air in slugs per cubic foot per second that is being affected and the acceleration is the change in downward air velocity in feet peer second from high above the rotor to far below it. The rotor induced power is thrust times the induced velocity at the disc in foot-pounds per second. This can be converted to horsepower by dividing by 550. Airplane aerodynamicists use "span-loading" in pounds per foot of wing span for a similar calculation of induced drag in forward flight.

A plot of "power loading" in thrust per horsepower versus disc loading can be made.

The top line is from the calculation for induced power. It represents only the power that is required to produce thrust. That is, of course, not all the required power to spin the rotor. The skin-friction of the blades must also be considered. This is done on this plot by other lines representing different "Figures of Merit." These are like levels of efficiency. The top line could be considered as representing 100% efficiency and would represent an unachievable rotor whose blades have no skin friction. The next line down at a Figure of Merit of 0.80 represents a rotor requiring 80% of the engine power to generate thrust and 20% to account for skin friction. The three helicopters are arbitrarily plotted at a Figure of Merit of 0.6 because the requirements for good forward flight performance motivate the designer to include more blade area than is optimum for hover.

We have methods for calculating the Figure of Merit for a specific rotor or it can be measured on a whirl tower. During preliminary design, the above plot could be used. For instance, your brother-in-law tells you that he met a man who is designing a new helicopter and needs financial support. He has decided it will weigh 10,000 pounds and will have a disc loading of ten pounds per square foot. He has a souped-up Cadillac engine that produces 500 horsepower so the power loading is 20 pounds per horsepower. You look at where that point falls on the graph and decide not to invest. You can give him advice, however, such as using two engines to give a power loading of ten. Or keeping the single engine and increasing rotor diameter from 36 feet to 86 to get a disc loading of only three pounds per square foot like the R-22.

Our experienced airplane aerodynamicist is facing more helicopter phenomena that don't jibe with what he has been working with.

Non-dimensional coefficients:

Whether when working with the forces and moments on a wing or a rotor, the aerodynamicist finds it convenient to use non-dimensional coefficients by dividing them by physical factors and flight conditions. For instance, the airplane aerodynamicist produces "lift coefficient" (C_L) by dividing wing lift, in pounds, by the dynamic pressure of the forward flight speed, $\left(\dfrac{\rho}{2 \times V^2}\right)$, in pounds per square foot, and the wing area in square feet.

$$C_L = \frac{\text{Lift}}{\frac{1}{2} \times \rho \times V^2 \times S_W}$$

Autogyros and helicopters have a similar coefficient obtained by dividing thrust by air density, in slugs per cubic foot (ρ), tip speed, ΩR, in feet per second squared, and disc area, in square feet.

$$C_T = \frac{\text{Thrust}}{\rho \times (\Omega R)^2 \times A}$$

Non-dimensional coefficients:

The tip speed is not based on RPM, but on Ω, in radians per second. Multiplying this by the radius of the disc gives the tip speed in feet per second. This is an important parameter that is selected during Preliminary Design.

To the airplane aerodynamicist, the dynamic pressure due to forward speed is a very important parameter, but to a helicopter aerodynamicist, the dynamic pressure at the blade tip that would be represented by $\dfrac{\rho}{2(\Omega R)^2}$, has no real significance and so the 2 is not included in order to have simpler equations.

The airplane also has drag (C_D) and moment (C_M) coefficients while the helicopter has torque (C_Q) and power (C_P) coefficients.

$$C_Q = \frac{\text{Torque}}{\rho \times A \times (\Omega R)^2} \qquad C_p = \frac{\text{Power}}{\rho \times A \times (\Omega R)^3}$$

A convenient relationship between these two coefficients is that they are numerically equal since: Power = Torque times Ω, in foot pounds per second. Data from rotor whirl tower tests are often plotted as C_P versus C_T. The equation for C_T can be used to estimate the disc loading for these tests by noting that if the tests had been done at sea level where air density (ρ) is 0.002378 slugs per cubic foot (a slug is a pound divided by the gravitational constant of 32.2) and if the tip speed had been 650 ft/sec, the first part of the denominator in the equation for C_T would be 1000. Thus for this situation, the disc loading would be 1000 times the plotted thrust coefficient. Of course, most whirl tower tests are not run at precisely these conditions, but it is an useful approximation.

There is another set of non-dimensional coefficients that has usefulness in helicopter studies. Whereas the first set was based on disc area, A, the new set is based on blade area, A_b, which makes them similar to the airplane coefficients. First, we have to identify another non-dimensional coefficient, σ. This is the rotor's "solidity", or the ratio of the blade area to the disc area, referred to as "sigma."

The new thrust coefficient is: $\dfrac{C_T}{\sigma} = \dfrac{\text{Thrust}}{\rho \times A_b \times (\Omega R)^2}$ in square feet

This is the "blade-loading coefficient", or in conversation, "Cee-Tee over Sigma." It can be shown that if the lift coefficient, C_l, is a constant for all blade elements, then: $\dfrac{\rho \times C_T}{\sigma} = \dfrac{C_L}{\sigma}$

This gives an opportunity to establish the maximum value of $\dfrac{C_T}{\sigma}$ by knowing the maximum value of C_l. For airfoils used on modern rotors, a maximum value of 1.2 is appropriate--giving a maximum value for $\dfrac{C_T}{\sigma}$ of 0.20. Any value above this represents a stalled rotor. (This was derived for a hovering helicopter, but it also applies to one in forward flight.

Of course, you don't want a helicopter that flies at the verge of stall so you choose parameters that ensure that it does not by specifying a moderate value of operational $\frac{C_T}{\sigma}$ such as 0.08. If the tip speed, ΩR, in feet per second, has been defined, then the blade area

is: $A_b = \dfrac{\text{Gross Weght}}{\dfrac{C_T}{\sigma} \times \rho \times (\Omega R)^2}$

The number of blades to make this area is "a designer's choice."

The blade loading coefficient is also important in determining the hover Figure of Merit.

Blade element theory;

The experienced airplane aerodynamicist is now faced with determining the performance of a helicopter in forward flight. He knows that if he were back working with an airplane that every wing element in forward flight would be subjected to the same velocity. This is obviously not true for a rotary-wing aircraft as shown below/

Distribution of Velocities

On first glance, this appears to be an impossible situation for such an aircraft because of a lift-induced rolling moment. The solution was found by Jaun de la Cierva, the Spanish inventor of the Autogyro, by attaching the blades to the hub with "Flapping Hinges." These make a blade's flapping natural frequency equal to the rotational frequency, (resulting in a system in resonance) and forces the blades to flap in such a way as to reduce the angles of attack on the advancing side while increasing them on the retreating side--just enough to eliminate the expected rolling moment. Modern helicopters also use cyclic pitch to do the same thing.

Reverse flow:

One feature of the velocity distribution is the "Reverse Flow Region" where the air is striking the trailing edge of the airfoil instead of its leading edge. As might be expected, this adds to the complications of analyzing the aerodynamics of forward flight because the blade elements in this region are stalled and are also producing negative lift.

Tip speed ratio:

It is time to introduce yet another non-dimensional term—the "Tip-Speed Ratio" which is the forward flight speed divided by the rotor tip speed, both in feet per second. As with other factors, it is given a Greek symbol, "μ"(mu). The reverse flow region will extend out the retreating blade a distance mu times the radius. For "pure" helicopters (without propellers, ducted fans, or jet engines) the speed is limited at a mu of about 0.5 due to retreating blade stall.

The next figure shows the angle of attack distribution for my example helicopter trimmed out at a tip speed ratio of 0.3 (115 knots) at sea level. The angles of 9 degrees in the third quarter are well below airfoil stall, but they indicate where retreating blade stall will occur at higher speeds or higher altitudes, or while trying to develop a high load factor.

FIGURE 1. Distribution of angle of attack

Choosing a tip speed:

The rotor tip speed is chosen during Preliminary Design based on both noise and high speed performance. Guidance to the designer is shown next.

Limits on tip speed:

Most designers will choose a tip speed of about 750 feet per second--or less--to minimize noise. The bottom horizontal line applies to the problem of rapid RPM loss in case of an engine failure. The stall limit line is plotted at a tip speed ratio of 0.5.

The compressibility line represents the condition where the Mach Number of the advancing tip is 0.92. Above that value, shockwaves make noise, increase drag, and make pitching moments that twist the blade—all undesirable effects. The plot indicates that the maximum speed of a pure helicopter is limited to about 200 knots. In reality, the official speed record is 216.3 knots set with a Westland Lynx in 1986. At this speed, the tip speed ratio was 0.5, but the advancing tip Mach Number was 1.0. (I was told that the flight was very noisy.)

Load factor:

The airplane aerodynamicist who is becoming a helicopter aerodynamicist has a surprise ahead of him.

An airplane pilot flying in level flight at a load factor of 1 can pull back on the stick and get a much higher maneuvering load factor. The faster he is going, the higher will be the change in load factor.

This is not true with the helicopter. The advancing blade, where the velocity is higher than that due to the hover tip speed, cannot take advantage of the potential increase in lift because of the requirement to minimize the rolling moment. The retreating blade, of course, is of no use in this maneuver. Thus it is up to the blades over the nose and tail boom to support the helicopter, but they think they are still in hover!

The difference between an airplane and a helicopter can be illustrated by looking at what happens when each aircraft encounters a gust.

FIGURE 2. Gust Encounter

As can be seen, the helicopter does not respond as much as the airplane. This has been demonstrated by formation flying comparable airplanes and helicopters in choppy air. The helicopters have a more comfortable ride.

The same effect occurs when trying to develop a high load factor by maneuvering in forward flight. Airplanes can be designed to develop maximum load factors from 3 to 6 depending on their requirements, but helicopters are generally limited to about 2.5 at the most. This can be traced back to the designer's choice of designing the rotor to operate in both hover and level flight at a $\frac{C_T}{\sigma}$ of about 0.08 while its maximum value is about 0.2.

Planform:

The airplane aerodynamicist may have shaped a wing's planform to minimize its induced drag. The reason is that a lift distribution in the shape of an ellipse is optimum by producing a constant downwash velocity in the wake. That this can be done with an elliptical wing as was illustrated by the British Spitfire fighter of World War II. Since such a wing is complicated to build, many wings are simply tapered to approximate an elliptical shape. Twist or "washout" may be used to minimize tip stall.

The design of the helicopter rotor for hover follows similar guidance. It can be shown that to get a high Figure of Merit, the induced velocity through the rotor disc should be constant. This can be approached by twisting the blades. A special twist that will produce a constant induced velocity over the entire disc and a high Figure of Merit is "ideal twist" obtained by changing the pitch at each blade element by dividing the pitch at the tip by the blade station of the element. (A similar effect could be achieved by increasing chord from tip to center of rotation in a similar manner.)

Figure of Merit

During his work as an airplane aerodynamicist, our new helicopter aerodynamicist had to predict propeller performance. You might think that since a propeller and a rotor are close cousins, the analysis of each must be nearly the same. Wrong!

Advance Ratio:

The Wright Brothers and other pioneers based their propeller analysis on the methods used for marine propellers. One parameter still used in propeller analysis is "advance ratio", known as "J." It is the distance the ship—or airplane—goes in one propeller rotation, divided by the propeller diameter. Its equation is: $J = \frac{V}{n \times D}$

Where V is velocity in ft/sec, n is the number of rotations per second, and D is the propeller's diameter in feet. Many propeller performance charts use J in one of their axes. When people in England began to study autogyros, they used a clean piece of paper. Instead of using "J" as an indication of speed, they selected "tip-speed ratio "(mu) instead. This is the non-dimensional ratio of forward speed to rotor tip speed. The two speed parameters are related—The propeller advance ratio, J, being the rotor tip speed ratio, μ, times pi. I think you will agree with me that μ is easier to visualize.

Activity factor:

Another parameter that the airplane people inherited from their marine brothers is the "activity factor." Each blade on most marine propellers is roughly in the shape of a disc. To account for the change in chord from root to tip and its effect on the capacity of each blade element to absorb power, integral calculus is used to represent a non-dimensional parameter,

$$AF. A_f = \text{No. of Blades} \times \frac{10,000}{16} \times \int_0^1 \left(\frac{c}{D}\right) \times \left(\frac{r}{R}\right) \times d(x)$$

Where c is the local chord, D is the diameter, and (r/R) is the blade station,. (I can't tell you where the 100,000 and 16 came from.)

Airplane propellers, of course, do not have disc-shaped blades, but the equation is used for their blade shapes

Activity factor is a complicated way of talking about rotor solidity, σ. For a propeller rotor with constant-chord blades, its solidity would be: $\sigma = \dfrac{128 \times b \times (A_F)}{(100,000 \times \pi)}$

Obviously, the helicopter definition of σ, as blade area over disc area, is a simpler way of getting this parameter

The original autogyros and helicopters did not have constant-chord blades. Their blades were made like a tapered sail-plane wings with spars and many ribs. In order to account for their shape a similar equation was used to calculate their "effective solidity".

We should be grateful to those early rotary-wing pioneers that we don't have to use propeller technology to do rotor analysis.

CHAPTER 39　　　　*The Windmill Brake State*

When discussing the vertical helicopter operating states, the "windmill-brake" state is usually the last to be considered. This is where the vertical rate of descent is so high that the rotor is absorbing energy from the passing air. It is something that can be simulated in a wind tunnel, but shouldn't be met in flight.

Windmills, as we know them, use that energy to do something useful, like pumping water, making electricity, or grinding corn. When simulating the condition in a wind tunnel, the electric motor that keeps the rotor turning for those other states such as hover and vertical climb is converted from a motor to a generator producing electricity. In most set-ups, the electricity is dissipated by running it through heating coils in a nearby room.

A helicopter in the windmill-brake state is converting potential energy into kinetic energy, but it. has no convenient way of using that energy except by increasing rotor speed, and there has to be a limit to that because of the high centrifugal loads that could tear the blades off.

This applies not only to vertical descent, but to descent at forward speed higher than that at which autorotation occurs. The region that should be avoided for my example helicopter is illustrated below.

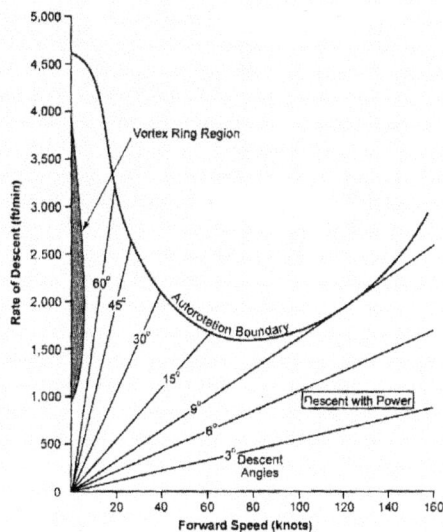

FIGURE 1. Windmill–Brake Region to Avoid

CHAPTER 40 *Another Use for the Tail Rotor*

I can be shown how a tail rotor at high forward speeds will flap enough to imitate an Autogyro rotor operating as a windmill. In this condition, it will deliver torque to the transmission just as the engine does.

It occurred to me that this odd behavior might be useful in another helicopter situation: to maintain rotor RPM in autorotation as an auxiliary power source after an engine failure. This could be done by side-slipping the helicopter to the right to simulate the tail rotor flapping that takes place in high-speed forward flight.

Using my text-book helicopter in autorotation at 80 knots as an example, it is getting its power by changing potential energy to kinetic energy at nearly the same rate as the engine produced in level flight at 80 knots before it quit: roughly 1000 horsepower. According to my calculations, the tail rotor in a right side-slip of 45 degrees will add about 40 horsepower to the system.

I'll have to admit that I had hoped for more, but 40 horsepower may be just enough to prevent an accident due to rotor speed decay.